Freedom:

Our last stand.

How can We Protect Our Freedom from the Immediate Attack from Feudalism, Indoctrination, and a Centrist Technocratic Tyranny?

The book summarises the 2020s period of post capitalism and how can we protect our freedom from the immediate attack coming from weaponised language, feudalism, indoctrination, and technocratic tyranny.

In Gates opinion “bioterrorism is the next big threat” and “we don’t worry enough” yet the vaccines themselves are a serious attack of this kind.

Unfortunately, Gates predictions have been miraculously accurate, so one must take them very seriously!

It seems clear that we are involved in a multistage process and Covid 19 was only one phase and protecting health is only a pretext.

Without a soul there is no sense of self, no memory or anything and no chance of recovery, you just exist like an empty shell.

“It is one of the most difficult things that can be imposed on a thinking person to have witness the course of a historical process among ignorant people, the inevitable outcome of which he has long known with clarity. The time of other people’s mistakes, of false hopes, of blindly committed mistakes becomes very long” Carl J. Burckhardt

The top-level connections set out in first 2 books is followed by 4 books that highlight the complex, social, economic, cultural, and technological issues for us to fix and a proven solution:

The Big 5 and globalists are taking over our tech world and getting greedier.

Our economies are being trashed.

And the sinister irony of an EV-only, 15 Minute City future, One World Government creeps ever closer.

Can we do anything about it?

Find out in this pragmatic, insightful and edifying book.

It’s time to look deeper.

But most importantly, we need to realise we’re being played like puppets.

When discussion stops, wars start. So, let’s join now to work out a solution to avert the weaponised dialogue and prevent conflict.

The purpose of this book is to set the scene and provoke thought, to fuel, inform and continue this much-needed executive debate.

Its urgent: we have a narrow window between now and 2030. Right now, 70% of corporate leaders are now unthinking and have become dope on a rope, so are inadvertently fuelling these loopy ideologically fuelled strategies.

Most leaders are naively and inadvertently trashing their businesses, yet they have no idea what is driving them.

This text describes the drivers, mechanisms that cause indoctrination, how we can escape and how we can regain our freedom.

How to make sense of madness: conspiracy theory or hypothesis?

We now face an escalating threat to humanity, that most are completely unaware of, and it seems that we are intentionally being made unaware of.

Ottawa Police detective Helen Grus was suspended for investigating whether parents of infants who had suddenly died were injected with mRNA products. Helen's concerns were scientifically based. A study by Hanna et al. demonstrated that mRNA injected mothers have S protein-producing mRNA in their breast milk. Her concerns, grounded in scientific research, should not be viewed as 'misconduct.' Instead, she was performing diligent detective work, exploring avenues others might overlook. McCullough Foundation 28-5-2024

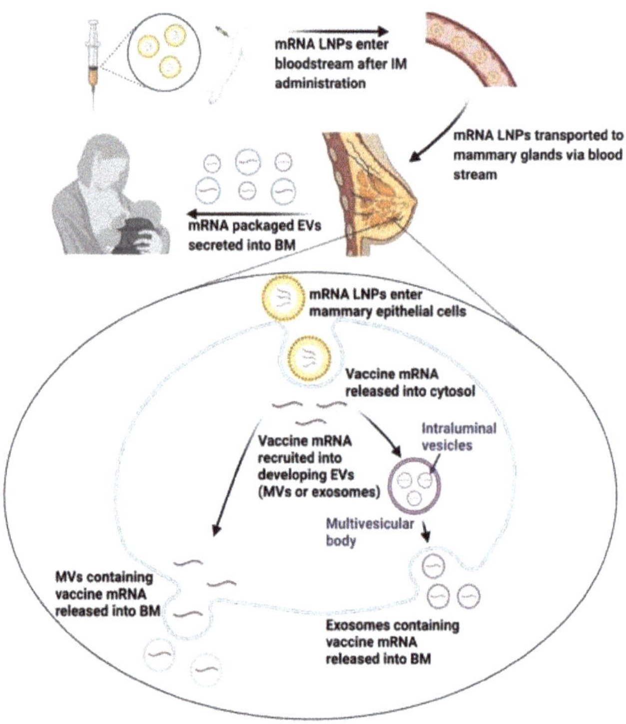

Source McCullough Foundation 28-5-2024

The big picture is complex. So, this book tracks to the best of my ability the outcome we are heading towards, the links with the science and economics, and the psychology driving the power crazed individuals that are putting us all at threat and how this strategy is unfolding.

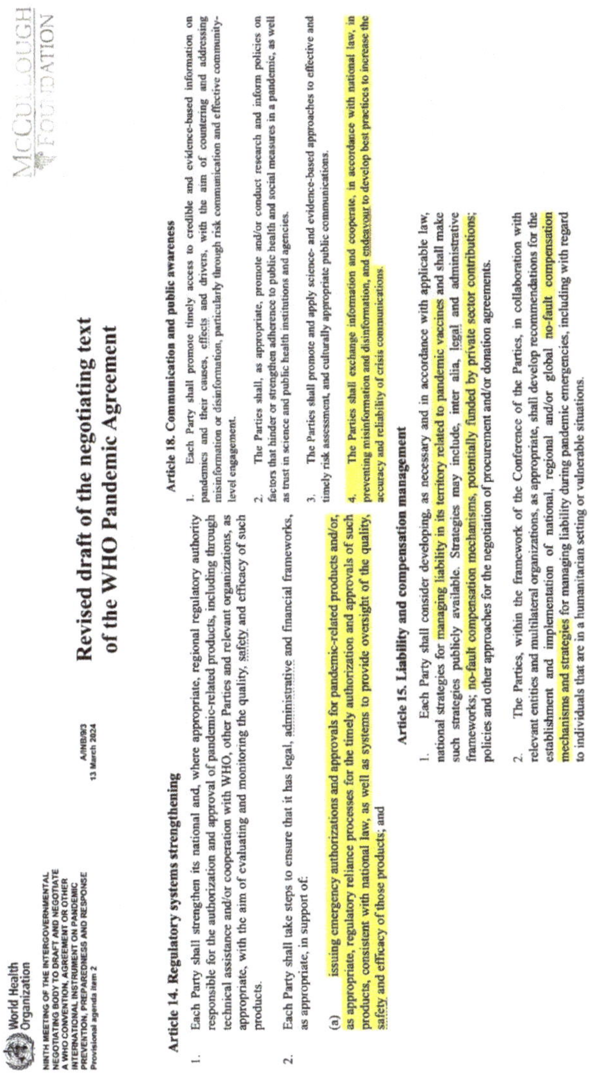

World Health Organization

NINTH MEETING OF THE INTERGOVERNMENTAL NEGOTIATING BODY TO DRAFT AND NEGOTIATE A WHO CONVENTION, AGREEMENT OR OTHER INTERNATIONAL INSTRUMENT ON PANDEMIC PREVENTION, PREPAREDNESS AND RESPONSE
Provisional agenda item 2

A/INB/9/3
13 March 2024

McCULLOUGH FOUNDATION

Revised draft of the negotiating text of the WHO Pandemic Agreement

Article 14. Regulatory systems strengthening

1. Each Party shall strengthen its national and, where appropriate, regional regulatory authority responsible for the authorization and approval of pandemic-related products, including through technical assistance and/or cooperation with WHO, other Parties and relevant organizations, as appropriate, with the aim of evaluating and monitoring the quality, safety and efficacy of such products.

2. Each Party shall take steps to ensure that it has legal, administrative and financial frameworks, as appropriate, in support of:

(a) issuing emergency authorizations and approvals for pandemic-related products and/or, as appropriate, regulatory reliance processes for the timely authorization and approvals of such products, consistent with national law, as well as systems to provide oversight of the quality, safety and efficacy of those products; and

Article 18. Communication and public awareness

1. Each Party shall promote timely access to credible and evidence-based information on pandemics and their causes, effects and drivers, with the aim of countering and addressing misinformation or disinformation, particularly through risk communication and effective community-level engagement.

2. The Parties shall, as appropriate, promote and/or conduct research and inform policies on factors that hinder or strengthen adherence to public health and social measures in a pandemic, as well as trust in science and public health institutions and agencies.

3. The Parties shall promote and apply science- and evidence-based approaches to effective and timely risk assessment, and culturally appropriate public communications.

4. The Parties shall exchange information and cooperate, in accordance with national law, in preventing misinformation and disinformation, and endeavour to develop best practices to increase the accuracy and reliability of crisis communications.

Article 15. Liability and compensation management

1. Each Party shall consider developing, as necessary and in accordance with applicable law, national strategies for managing liability in its territory related to pandemic vaccines and shall make such strategies publicly available. Strategies may include, inter alia, legal and administrative frameworks; no-fault compensation mechanisms, potentially funded by private sector contributions; policies and other approaches for the negotiation of procurement and/or donation agreements.

2. The Parties, within the framework of the Conference of the Parties, in collaboration with relevant entities and multilateral organizations, as appropriate, shall develop recommendations for the establishment and implementation of national, regional and/or global no-fault compensation mechanisms and strategies for managing liability during pandemic emergencies, including with regard to individuals that are in a humanitarian setting or vulnerable situations.

Source: McCullough Foundation 2024

Speaking at the Copenhagen Democracy Summit earlier this month, President of the European Commission Ursula von der Leyen suggested the audience think of information manipulation as a virus and then said it's better to vaccinate than treat the infection. "Think of information manipulation as a virus," she said. "Instead of treating an infection once it has taken hold, that's the de-bunking, it is better to vaccinate so that our body is inoculated. Pre-bunking is the same approach. Because disinformation relies on people passing it on to others – it is essential that people know what malign information's influence is and what the techniques look like."

How severe is the issue? The Nurenberg Treaty was created post WW2 to prevent the likes of the Nazi, Dr Mengele imposing medical trials upon people against their will. Yet it seems that there is justification to believe that the centrist entities of the World Economic Forum, UN and World Health Organisation are complicit in the same evil legacy all over again by imposing their unproven medical trials upon us.

To counter and test an irrational threat in the face of attempts to silence us, we need to be able to maintain and refine sound hypotheses, as well as making sense of the psychology to develop and refine conspiracy theories.

A brief introduction to the narrative driving the Pharma industry –

Attempts to silence indicate an attempt of a perpetrator to hide from the truth. So, to make sense of the combined threat of silencing, and the obfuscation of analysis, we need to

1. develop a hypothesis as a basis to design experiments or to collect data,

alongside...

2. a conspiracy theory, a belief that some secret but influential organization is responsible for an event or phenomenon.

"Cancelling Hypotheses (Conspiracy Theory) Description: The argument defends one hypothesis by proposing a second hypothesis to explain the lack of evidence in support of the first hypothesis. That is, the second hypothesis cancels or undermines the predictions made by the first hypothesis. Conspiracy theories are a particularly important example of this fallacy. The conspiracy generally involves a conscious effort to destroy or cover up the evidence that would otherwise have supported a given hypothesis. However, the term "conspiracy theory" is generally used only when there has been a conscious effort to hide evidence, and not all examples of Cancelling Hypotheses involve such conscious effort." (Source Palomar Edu)

The conspiracy theorist is therefore the perpetrator not the whistle-blower.

The fraudster will create a chain of energised generalisations or statements, on diverse topics such as climate change to gender to mass vaccination. They fire these in quick succession to dupe their audience, to push their own ill founded, unchallenged ideological agenda. Many corporate leaders unwittingly use this tool. Leaving their subordinates with little choice but to follow and not speak up.

Understanding the psychology of a scenario, and the consistency of the psychology with the actions taken, is a means of making sense of irrational agendas. Or put it another way, understanding the underlying thinking is vital to make sense of a scenario, because it is our thinking whether rational or irrational; that drives everything we do.

These individuals have a myopic focus to control us to realise their 2030 dream and New World Order.

Rockefeller decades ago, realised that his grip on the oil monopoly was wearing thin, so he decided to launch oil-based medicines, and did all he could to crush natural products. Over the last 40 years, the investment in oil has switched to technology and the exponential growth and power of the Pharma industry and artificial intelligence.

This investment focus has created the dominance of the current Pharmacology industry, along with the legacy of the expertise of discrediting of all alternatives: be it natural products and scientists that don't follow their dominant mantras. Yet from my experience of the pharma industry largely, is that the body naturally fixes itself.

The dominant globalist mantras corrupt. These mantras have led to the launch, normalisation of deadly mRNA “vaccines” by making every effort to discredit alternatives. Thos continues despite mRNA leaving a substantially harmful legacy of cognitive and long-term damage to health to the 4Bn people that took them. The prions from these spikes, may now be impossible to eradicate from of the human population and environment, making us dependent on the next stage of treatments to reverse the damage from the first spike.

To make avoiding ingesting mRNA impossible, there are now reports of the global elite spiking 100m animals with mRNA. Dr Kevin McCairn warned that these prions will now spread between species and into the land and back to us so we can’t now escape the mRNA threat.

Dr Kevin McCairn became a “discredited scientist” when he shared his concerns about the impact of these gene modifying technologies on the brain and organs. In 2020 he raised the flag about the dangers of the contagion of prions to the heart and brain and blood vessels. These prions replicate like Lego into insoluble blood clots, that we have no protection against

Dr Nehls too raised concerns about the mechanism of the role that mRNA vaccines play in indoctrinating people by reducing hippocampus function thereby causing dialogue to overwrite memory and effectively rewrite our souls.

There is a hope because for example about 20% of Germany have NOT been vaccinated, although nearly 100% of Israelis tragically were vaccinated and much of Africa weren’t vaccinated, but they may present less of an economic threat, without the same resources to resist the globalists plan…

Yet this nefarious “show” goes on, this book explains why and how… and what we can do to get a grip.

Contents

Post Capitalism &
The Feudal
Mousetrap
We must join together to avert tech-fuelled feudalism, China's plans, digitisation and automation, and realise we're being psychologically played like puppets.
Tom Pickering
Post Capitalism & The Feudal Mousetrap
The Big 5 and globalists are taking over our tech world and getting greedier.
Our economies are being trashed.
And the sinister irony of an EV-only future creeps ever closer.
Can we do anything about it?
Find out in this pragmatic, insightful and edifying book. It's time to look deeper into AI and its impact on us meagre humans. We must discuss tech-fuelled feudalism, China's plans to monopolise the world, and digitisation and the ramifications of anti-labour automation.
But most importantly, we need to realise we're being psychologically played like puppets.
When discussion stops, wars start. So let's join together now to work out a solution to this madness before the woke-fuelled tyrannies infiltrate any more unsuspecting minds.
Let's get to grips with the psychology driving this to inform this much-needed debate.

The Big 5 and globalists are taking over our tech world and getting greedier.

Our economies are being trashed.

And the sinister irony of an EV-only future creeps ever closer.

Can we do anything about it?

Find out in this pragmatic, insightful and edifying book.

It's time to look deeper into AI and its impact on us meagre humans. We must discuss tech-fuelled feudalism, China's plans to monopolise the world, and digitisation and the ramifications of anti-labour automation.

But most importantly, we need to realise we're being played like puppets.

When discussion stops, wars start. So let's join together now to work out a solution to this madness before the woke-fuelled tyrannies infiltrate any more unsuspecting minds.

Let's continue this much-needed debate.

“Logic can be met with logic, while illogic cannot-it confuses those who think straight. The Big Lie and monotonously repeated nonsense have more emotional appeal in a cold war than logic and reason. While the enemy is still searching for a reasonable counterargument to the first lie, the totalitarians can assault him with another.” — Joost A.M. Meerloo

Why Can’t We Stop? We’re under threat. A threat that, left unnoticed, will culminate in mass hysteria. It’s time to bridge the gap and question this threat of menticide because if left to seep into our subconscious and we don’t understand how to tackle it, our voices will diminish. One by one. From the Bible and best practice to hundreds of company turnarounds, psychological analysis and a summary of the strategic shifts of the 2020-2021 crisis, the author draws on an eclectic range of sources to summarise his message for further consideration. The intention of this book is not to brainwash you into the author’s way of thinking. Rather, the purpose of this pragmatic and practical book is to highlight the dire crisis society faces, encourage intelligent debate, and provide you, the reader, with the knowledge to take back control. So what are we facing? Are people simply becoming increasingly stupid? Or is tech getting cleverer? Are the boundaries between logic and illogic becoming so blurred we can’t think straight? How do can we handle the tyrannical views and shallow ideologies that swamp the media? Are we looking high enough up for our answers? Let’s start the debate.

Regardless of your business or personal scenario, this book shares how to realise "your infinite capability," and "regain what good looks like."

"You have infinite ability" How to get back to what good looks like. The antithesis of our current hell. In the 1930s there were very few people attending Hitler's Nazi party rallies who saw that there was something wrong, and so chose not to salute. What did those people see that the rest did not and why are people also now silent? In the aftermath of the Covid 19 Pandemic, we now face the emergence of a similar loss of purpose seeding woke culture fuelled by tech-driven tyrannies. This book shares the tenet and methodology that shines a light into the new perspectives to realise and "regain what good looks like." Tom is an award-winning business leader and company turnaround specialist who openly shares his cumulative business turnaround experience drawn from how to achieve immediate results. These scenarios present human behaviours at their most entrenched. To realise the benefit from this book you must take new action. Each chapter provides you with the catalyst for a new observation, a self-realised new action, and the learning of having taken that new action. This book is a truly exciting journey for those that find the courage to bring with them an open mind. Winningthinking® breaks through barriers and enhances cognitive strategy to accelerate the process of transformation in people, teams & organisations.

What does good look like?

I have found this is what a good business looks like - a jazz band. A jazz band has an intimate connection between its players. They have fun, rely on each other, and observe the audience to keep their music wholly aligned with a successful outcome. The human connection and intimacy with the audience are enthralling.

AI-created music has no soul and manipulates our experience by manipulating our human dopamine and serotonin systems. This tech manipulation, like alcohol or heroin, creates addiction.

“And no one could buy or sell anything without that mark, which was either the name of the beast or the number representing his name.” NLT

Who is in control? Does the narrative control us, or do we control the narrative? Who is in control and how can we get control?

We are creating a society where the young are switching to aggressive language or physical violence and stabbing before and instead of conversation to resolve increasingly trivial issues. China’s young adults (the WEF’s digital panacea) who cannot find work and are forced to live with their parents are being classified and accepted as “full-time children”.

The speed of human social development is unprecedented.

So, what is causing this? What are the historical precedents? How can we get a grip? And what can WE do about this?

It’s a psychological challenge.

Chinese firms are driving for people-less facilities.

The tech Big 5: Alphabet, Amazon, Meta, Microsoft and Apple keep growing. So far in 2023, their valuations have soared by half to around $9Tn, or a quarter of the value of the S&P500. They account for 60% of the R&D spend of all the firms in the index, so they are likely to define the strategy and be the main winners in the “tech revolution.”

Governments view this domination with trepidation and try to enforce antitrust measures. They are so big that they are sucking the life out of challengers. Garmin is one of the largest challengers in the smartwatch niche with net earnings of $974m, which is a drop in the ocean against these Big 5.

China attack at a time of their choosing, and with phase 5 of their strategy underway (to export their own products), China is targeting Apple, the US's most valuable company, catching up on microchip design and manufacture, as well as beginning to (if not dominating) the OEM electric vehicle (EV) design manufacture and the EV material supply chain.

> A technological advantage in A.I. would create the risk of unstoppable cyber attacks — a taste of this was delivered recently by Chinese hackers, targeting U.S. financial services and utilities. More advanced A.I. would give China more capacity for censorship, amplify their naval power, and enable greater persecution of political dissidents.
>
> onezero.medium.com/the-future-of-a-i-is-probably-chinese-

The export of China's products and influence comes with the inherent psychology that underlies their technology. Western governments seem unable to avoid getting drawn into the same surveillance and vast data-led obsession. Hence, this influence comes with nefarious Chinese controlling ideology for fear of being left behind in what may be a Chinese-led race to the bottom.

The psychology of China:

"National Security Strategy," published in December 2017.

"For decades, U.S. policy was rooted in the belief that support for China's rise and for its integration into the post-war international order would liberalize China. Contrary to our hopes, China expanded its power at the expense of the sovereignty of others," the document declared. "We will work with our partners to contest China's unfair trade and economic practices and restrict its acquisition of sensitive technologies."

medium.com/bloomberg/another-reason-u-s-fears-huawei-its-gear-works-and-its-cheap

China has a long-term technology strategy; one strand of it is Chinese pure battery cars, which now make up 8% of EU sales. UBS estimates that 20% of all cars sold in Europe could be Chinese made by 2030. China plans to make 50% of all global cars, and claims of anti-trust based upon Chinese Government support look hypocritical.

The crux of the administration's argument is that Huawei couldn't say no to demands from China's government, even if it wanted to. China's National Intelligence Law of 2017 requires all Chinese companies and citizens to "assist in and cooperate in national intelligence work" if requested. Whether Huawei or any of its employees have done so in the past is now incidental, China hawks argue, given Huawei's lead in developing 5G technologies for the future.

bloomberg/another-reason-u-s-fears-huawei-its-gear-works-and-its-cheap-

The Chinese cost advantage is 25% due to the vertical control of their supply chain. Half of German car firms' profits come from China, and other budget marques face a daunting challenge in the face of the cutthroat in the increasingly commoditised mass market. The global electrification of motorbikes is moving much faster.

Getting strategy right is truly an intertwined global problem. So what is driving this? Can we get a grip of it? And what are the historical precedents to help us anticipate what might happen?

MIT Technology Review

Perhaps we just need to be patient; technology advances have always increased incomes, which then increased demand for goods and services, which then led to more jobs. But Laura Tyson, a top economic advisor to President Bill Clinton and a professor at the University of California, Berkeley, asks the question that is on everyone's mind: What if, this time around, the goods and services that people want just don't require much human labor to produce? "This is the first time that technology, we think, could on net reduce the demand for human workers," she says.

Greed and the curse of more

These 5 tech organisations now control 20% of the US GDP. Monopoly owners always have done all they can to protect their wealth and power.

People are naturally keen to emulate these technology leaders and build wealth, so they naturally follow and emulate them.

But at what cost?

People are being left behind.

Good jobs have primarily disappeared for those without a degree, and we are witnessing skyrocketing inequality. The post-WW2 period saw an increase in real incomes. No longer.

Digital technologies are everywhere and have made vast fortunes for their founders and a fraction of the population. Yet, real wages have hardly increased, and people without a college education have seen their earnings and opportunities decline.

There is nothing automatic about new technology bringing widespread prosperity. Indeed, Claus Schwab's mantra is exactly the opposite: the population in the great reset "will own nothing and be happy."

The various WEF climate, economic, anti-oil and other strategies are trashing our economies. For example, China is being allowed to run roughshod over our automotive and other key industries, leaving no residual interest in building indigenous value. Currently, 240 of the 320 giga plants globally producing EV batteries are Chinese owned.

These global shifts look unproven but, in fact, have all the traits of going back to a centralised feudal system alongside the economic rent that citizens have plagued civilisations for millennia.

Technology has upended the economics of modern businesses. Jonathan Taplin former rick band manager and movie producer points out that “musicians could once make a decent living off their royalty checks”. He sees what has happened in music as the harbinger of much of the rest of culture: “the rise of the digital giants is directly connected to the fall of the creative industries in our country”?

What determines when wages and new techniques increase wages?

The underlying ideology or psychology the New World Order, or the UN's 17 Sustainability Goals have always been misleading. Yet 179 countries have signed up to them. These goals are fuelled by misleading mantras such as "productivity increases" and the "green economy". Economic theory suggests that as productivity increases, corporations expand output, and wages increase because labour, through scarcity, increases their bargaining power.

There is much talk about accepting and adjusting to this change, as if it is inevitable and progress, so instead of racing against the machine, maybe we should race with the machine?

But productivity growth isn't translating into higher wages for workers. Even when sales per employee increases, whilst automation may increase average productivity, that may reduce output per employee.

So, whilst automation and offshoring may have raised productivity and multiplied corporate profits, nothing resembles an increase in shared prosperity in the West.

Even more important with new technology is the creation of new tasks. With the launch of tractors in farming, a whole new industry sprung up, creating many more and, in some cases, higher-skilled jobs for those displaced making these tractors.

But current technology is not creating such new industries. For example, EVs reduce the vehicle's complexity to a fraction vs internal combustion engine (ICE) vehicle, and they reduce the profit driver: the differential value adds an opportunity to innovate in the EV industry.

> The new industries were highly productive and also subject to furious technological advancement. Weavers lost their jobs to automated looms; secretaries lost their jobs to Microsoft Windows. But each new spin of the technological wheel, from plastic toys to televisions to computers, yielded higher incomes for workers and more sophisticated products and services for them to buy.
>
> Something different is going on in our current technological revolution. In a new study, David Autor of the Massachusetts Institute of Technology and Anna Salomons of Utrecht University found that over the last 40 years, jobs have fallen in every single industry that introduced technologies to enhance productivity.
>
> The only reason employment didn't fall across the entire economy is that other industries, with less productivity growth, picked up the slack. "The challenge is not the quantity of jobs," they wrote. "The challenge is the quality of jobs available to low- and medium-skill workers."
>
> Adair Turner, a senior fellow at the Institute for New Economic Thinking in London, argues that the economy today resembles what would have happened if farmers had spent their extra income from the use of tractors and combines on domestic servants. Productivity in domestic work does not grow quickly. As more and more workers were bumped out of agriculture into servitude, productivity growth across the economy would have stagnated.
>
> "Until a few years ago, I didn't think this was a very complicated subject: The Luddites were wrong, and the believers in technology and technological progress were right," Lawrence Summers, a former Treasury secretary and presidential economic adviser, said in a lecture at the National Bureau of Economic Research five years ago. "I'm not so completely certain now."

medium.com/the-new-york-times/tech-is-splitting-the-u-s-workforce-in-two-

But this simplified, less value-added model suits the Chinese mindset of domination, attack, and profitless growth. This space is where profitless growth strategies usurp Western EV players whose investors are, sensibly, profit driven. Furthermore, the value is in making the batteries. China controls the battery supply chain and fire its power stations with cheap coal to create highly polluting electricity, producing a vast amount of electrical energy to process the lithium to manufacture the batteries.

So, a broader economic perspective of value building in a macroeconomic global environment is now required to survive, protect our economies, and pay our bills. Yet it will only be an increase in local wealth that creates the funds to pay for this and from which everything else follows. Simply put, people without money don't care about the environment. It's not a priority.

The reason that Keynes feared technological redundancy and until recently it didn't come to pass, was because new industries created tasks for workers.

MIT Technology Review

Perhaps we just need to be patient; technology advances have always increased incomes, which then increased demand for goods and services, which then led to more jobs. But Laura Tyson, a top economic advisor to President Bill Clinton and a professor at the University of California, Berkeley, asks the question that is on everyone's mind: What if, this time around, the goods and services that people want just don't require much human labor to produce? "This is the first time that technology, we think, could on net reduce the demand for human workers," she says.

Most technology investment focuses on surveillance rather than productivity improvement or extracting more effort from workers, thereby reducing their value for the same or less pay.

This PWC model is based upon the defunct economic supply, demand, and pricing model, so hiring more workers is almost certain NOT to come to pass. This is because NOW, new industries and new jobs for people are NOT being created.

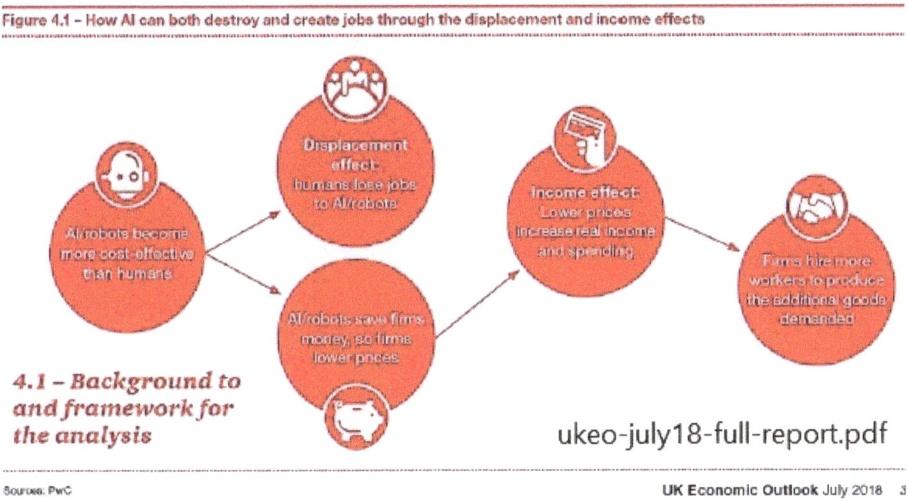

What is the Choice, Feudalism, or human wellbeing?

Technology is not inevitable; we have choices. We could use technology to increase automation or surveillance or create new tasks that benefit and enhance work and workers.

Surveillance-related applications are not unlike slavery: impoverishing and increasing coercion.

Historical precedents are clear: employers don't increase wages unless they face competition from rivals to do so. In many peasant communities, peasants in a feudal system were tied to the land so they couldn't seek employment elsewhere.

AmazonGo is an example of what new feudalism looks like:

In the U.S., the average convenience store visit lasts about three and a half minutes — including the walk from car to store.

I don't really care about saving 50 seconds, especially if it means subjecting myself to incredibly intense surveillance. What might Amazon do with all the behavioral data it collects during my lunchtime sandwich run?

In the UK in the Middle Ages, employees were prevented from working elsewhere and faced jail if they tried to take better jobs. Most of the spare output and abundant labour were utilised during the massive construction boom, building churches. So the factory owners became fabulously wealthy.

Wages are always negotiated by rent sharing rather than left to market forces. The vast wealth created by tech firms now equates in the same way to the same economic rent model. This wealth in the Middle Ages was well over and above the prevailing norms, akin to the wealth of the Big 5 tech companies' parasitic domination and the way the system feeds on productive aspects of the economy. The factories of the Middle Ages created useful products such as cloth, whilst the parasitic tech firms now create a feudal sales intermediator layer that takes the lion's share of the rent or profit between the consumer and the provider of the goods and or the services sold.

Economic rent – the $100Bn gap

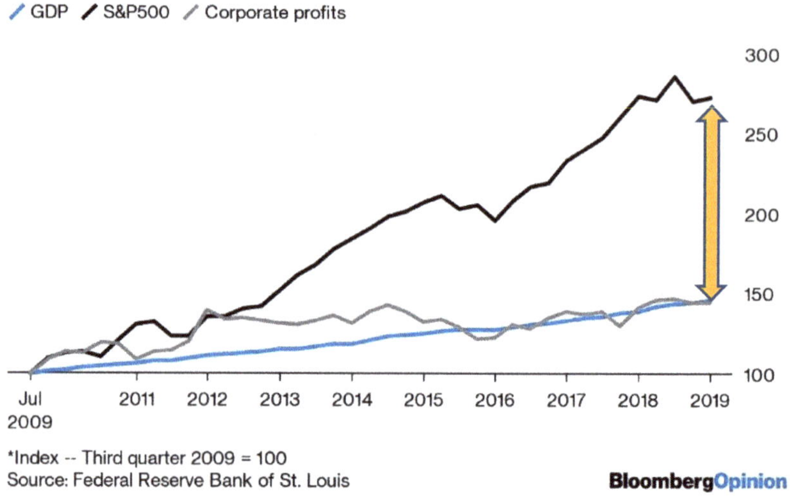

These decisions that determine who win and who loses from these new digital excesses should be made collectively by visionaries and political and corporate leaders.

But without some checks and balances, history is clear that it is almost certain NOT to happen.

Technology has revolutionised science and reduced barriers to sharing and spreading ideas. This has built cumulative value at lightning speed vs earlier innovations, especially when a pressing need exists. In the 1600s, the knowledge and learning of Isaac Newton and other innovators diffused very slowly. James Watt's early development of the steam engine was enabled through close collaboration, and the technology was leveraged and applied in different ways. So James created a shared vision and enabled approaching different problems in different ways.

Who wins and what role does democracy play?

Whoever has the greater power is most likely to persuade others to their perspective, which, of course, is aligned with protecting their interests, power, building, and wealth. The wake of their influence leaves others following blindly in their footsteps, hoping some wealth or success rubs off on them.

A shared vision or the assumptions underlying technology: surveillance, convenience, or efficiency, ESG objectives, societal control, productivity, or profit improvement locks strategy in a specific strategic direction. These assumptions fuel narratives to form the prevailing wisdom. This makes it difficult for employees, or even corporates, to break out of its hold, whether the hold is good or bad.

So, the direction of progress is socially and psychologically biased in favour of those with power and against those stakeholders without a voice, thereby trashing democracy or the population's ability to steer or influence strategy from the bottom up.

So, the different ways of organising technology and/or the associated ideology, assumptions or narratives intentionally empower some groups and disempower others.

The implications of quashing opposition to gain control and/or increasing revenues can happen on a vast scale. The Chinese Communist Party's (CCP) social credit system affects the lives of 1.4bn people. This scoring was a decision by very few party leaders without consultation.

In the same way, Facebook, whose underlying psychology/mantra is "move fast and break things", Zuckerberg introduced a policy to increase the engagement of their 2.4bn users with their friends. All this did was amplify polarisation and misinformation and take users' attention away from factual posts and newsfeeds. This too was implemented without consultation with users or government input.

Google's CEO said that artificial intelligence (AI) is "probably the most important thing anyone has ever worked on, more profound than electricity or fire". But countering such power with alternative sources of power, harnessing the fallout or selfishness, or creating a more inclusive vision would be desirable. Yet, it is the antithesis of the psychology of artificial intelligence.

So, AI is taking power away from people and concentrating the power and investment towards those controlling data and making key decisions. AI undermines democratic institutions, fuels inequality and strengthens autocracy. The underlying narrative surrounding AI is always focussed on the obsession with data and therefore an increase in surveillance and reduction of privacy.

Consider the psychology of climate change and/or fixing the narrative or fallout of the circular obsessive mantras of: "more technology will fix it" and "technology is inevitable".

Ray Kurzweil's psychology is that "with AI we will reach boundless prosperity". AI will surpass human capabilities, and/or human intelligence will merge technology with humans to create super AI humans. The current level of influence of technology via smartphones is the first stage of such transhumanism.

medium.com/@Michael_Spencer/fear-automated-future-robot-economy-984e3a309ba3

We think many in the bottom 40% will be fundamentally left behind. That is, they will live in a different era than the financial elite, with less access to the very institutions, knowledge and AI that would enable them to participate in the future of work. Artificial Intelligence and not the American Dream, will increasingly govern our lives and the future of our opportunities, as we are already seeing now.

We should be fearful and adapt to an increasingly automated future. Robots are coming, and then robots able to build smarter versions of themselves. Finally, Artificial Intelligence able to self-learn, at a pace homo sapiens are certainly incapable of matching.

To paraphrase Hassabis, he says, "either we need an exponential improvement in human behaviour, or we need an exponential improvement in technology." Yet what is clear is how little we know about our existing potential or capabilities and/or even what a human being is.

What is clear is that technology, contrary to the WEF mantra of societal control using technology as a good or necessary thing, is escalating societal violence by trashing communication skills. When speech is no longer an option, the only option left is verbal violence, embedded in the attack dialogue of wokism and/or escalating knife crime. Conflict is now triggered by increasingly trivial issues between young people, who seem to lack the skills to communicate or show grace or heart to resolve day-to-day matters.

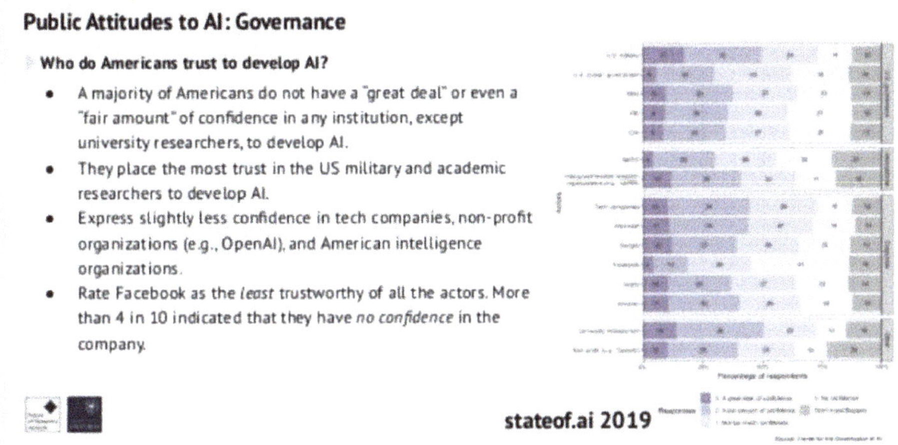

What we are witnessing is not progress towards a common good but a common vision shared amongst the technology leaders, all of whom think the same: surveillance and data collection without sharing prosperity. This tech elite is wrestling control from our democracies and governments.

Attempts to take control of government define fascism. The influence amplifies the tech elite's wealth and power at the expense of ordinary people by circumventing and influencing the government. The primary purpose of the state is simply to look after the poor or casualties of society who cannot look after themselves. These underlying tech elitist ideologies create blind spots that ruinously disregard the voice of the powerless individuals in society, control the media and narrative to meet their own myopic purpose: retaining their wealth and maintaining control, the basis of in conjunction with the WEF, their own central globally governed system.

Does the narrative control us, or do we control the narrative?

Holding people in a fearful state causes them to stop thinking and just do. Non-thinking people are easy to manipulate. This control is maintained by forcing escalating climate emergency or other compounded negative mantras that self-reinforce their ideologies. Ideology propagated by the WEF, consultancies, or ESG-driven elite corporates control the dialogue that misaligns actions, controls beliefs, and isolates dissent.

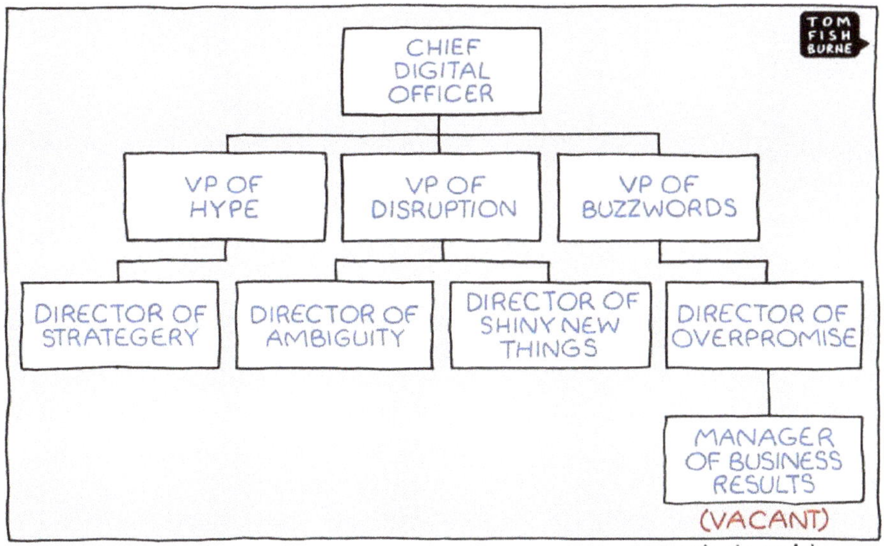

© marketoonist.com

The icing on the control to destroy someone's will is when you can get people to enact your own ideological falsehoods when the individual knows intuitively that your proposed actions and controlling mantras are wrong. Then you have complete control of them. This is the seedbed for menticide and driving change ever faster, without dissent in complete silence.

How Do Big Tech Giants Make Their Billions?

And in the 2021 fiscal year, big tech saw a **27%** growth in combined revenue, year-over-year.

Company	Revenue (FY 2020)	Revenue (FY 2021)	Growth (YoY)
Apple	$274.5 billion	$365.8 billion	33%
Amazon	$386.1 billion	$469.8 billion	22%
Alphabet	$182.5 billion	$257.6 billion	41%
Microsoft	$143.1 billion	$168.1 billion	17%
Meta	$86.0 billion	$117.9 billion	37%
Combined	**$1.1 trillion**	**$1.4 trillion**	**27%**

How did these companies continue to thrive throughout economic turmoil and global chaos? It was made possible because the societal changes triggered by COVID-19 ended up driving demand for big tech's products and services. https://www.visualcapitalist.com/how-big-tech-makes-their-billions-2022/

To create a successful outcome, the prevalent technology tool must work for people and challenge the worldview that prevails amongst the global tech elite whilst checking the assumptions, accepting, for example, that contrary to the tech elite mantras, there are no automatic productivity gains.

Technological transitions have always tended to enrich the elite whilst generating few benefits for the masses, and peasants lacked the political or social power to challenge them. Technology and the suppression of media and dialogue, addictive control of attention driven by the tech elite, provide a whole new level of tools to manipulate people, dominate their attention and trap them in a bubble

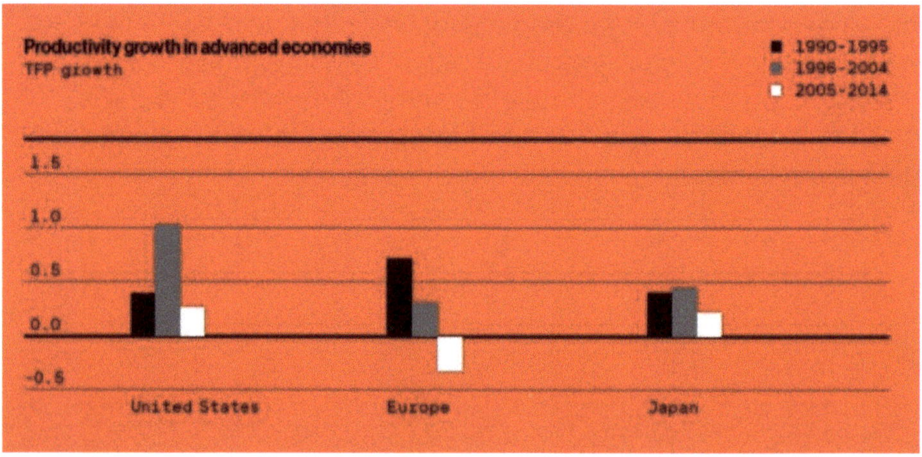

or echo chamber to undermine collective bargaining by atomising dissent and cancellation.

The Milgram experiment showed that circa 70% of people would knowingly do the wrong thing if told to do so. The concern is that by addicting people to the technology, as well as imposing false ideology upon them and giving them no option but to enact these falsehoods, you will trash the will of the hard core 30% that care. This hardcore includes the likes of Laurence Fox and Russel Brand. They both care, but by being provoked by the underlying lies, they and others who care, become vulnerable to behaving badly in response to the propaganda of the tech, gender, climate and or vaccine type ideology.

The first phase of the Industrial Revolution was disempowering for most people. It only started changing when people started organising to force economic and social change, akin to abandoning a shared prosperity model. In other words, a shared prosperity model has never emerged without social and political dissent and people revolting against the elite.

Technology is currently not focused on creating new opportunities and tasks for people but towards cost-cutting and cutting labour costs. Indeed, FANG companies, alongside Chinese titans like JD.com, are now creating a mega online business with the objective of employing no one.

Yet the wealthy masses have power, which conflicts with the tech titans' objectives to dominate and push back on input from workers, labour organisations and government regulation. So by disempowering people, AI is turbocharging this economic inequality, leaving ordinary working- or middle-class people disengaged, atomised and left behind.

None of this is inevitable. No one is asking how these machines can become useful to humans. At the same time, AI advertisement models amplify extremism.

Undermining democracy and what is good for the economy are the perfect seedbeds for the tech titans to maintain their control.

The power of a compelling vision

Once a compelling ideological (false) vision takes hold, say the escalating dialogue around "Climate emergency" or "technology inevitably driving progress and productivity", change takes place for good or bad. The power of the dialogue is difficult to speak up against for fear of being labelled and isolated as a climate change denier or a Luddite.

Leaders in these situations fail to admit to difficulties or the underlying falsehoods, even when these falsehoods become too obvious to everyone else, such as gender ideology, feminism, critical race theory, queer theory etc, all of which fuel mental health, male suicide, isolation, atomisation and loneliness.

In the case of building the Panama Canal, Ferdinand de Lesseps refused to admit to difficulties. This cost the lives of tens of thousands of people to no avail. Furthermore, such leaders need to be challenged by counter-views from those around them. But instead, hubris takes hold, blinding them and their peers to their stupidity. This fuels the blind spot that the tech elite are "men of genius" and deserve the highest social standing. They have earned the right to ignore others by uniquely knowing what is best for others, for the climate, or Xi's Chinese technological structure of civilisation. This leaves ordinary people with the psychology of the elite conveniently labelling the masses as stupid, idlers or lazy.

But this is to misunderstand where the real problem lies. It is not Artificial Intelligence that should terrify you, it is natural stupidity!"

Dr Nigel Shadbolt, professor of computer science at Oxford University.

Lesseps, who led the Panama Canal fiasco, justified his hubris with this psychology: “Remember if when you have something important to accomplish, that if there are two of you, there is one too many”. This is hardly the seedbed of democracy or sound, balanced decision-making.

So, Lesseps regarded the mass loss of life or, in the case of the tech titans, case addiction and manipulation, and mental illness as an acceptable cost of progress. The blind spots don’t allow learning from setbacks on the ground, such as setbacks and abject failure, technology, and crypto. To the contrary, AI development is accelerating to narrow the window of dissent. In addition and as evidence of the fascist control the WEF, WHO, tech titans, etc, have on government, military-grade propaganda was imposed by governments on our populations to accelerate the transition towards “the great reset” during and to silence those during the shutdown.

The left-wing Marxist ideology coming through schools used to be based upon control using the perceived systemic Marxist economic inequality victim-oppressor strategy. But the economic case lost its relevance in the 1950s, so the likes of John Money introduced a variation of the Marxist victim-centric group identities around race, gender, sexual mutilation, etc, to gain control. This ideology in its various forms, has been in our schools for three decades. Three decades is the time deployed by the Russian state to get the population to a manipulatable state, alongside imposing ideology and division so people become incapable of thinking for themselves. They become subservient to the state or, in this case, the tech/fascist central controlling governments.

Technology is nothing without vision. Yet, how distorted is the psychology of that vision in limiting what people can see or want to see? In the case of the Suez Canal, the Egyptian labourers building it played no part in its future or the success of its future. The technology elite have the same psychology as Lesseps: belief in investors, blind techno-optimism, silencing, and indifference to the fate of the voiceless.

In 1879, one delegate referring to Lesseps, said, “The Failure of Congress will teach the people the salutary lesson that people must think for themselves and NOT follow the lead of one man”. Has this lesson never been learned?

Social power

It is social power and a powerful network backed by a compelling story that enables people to convince thousands of others. “Social” media alongside AI is the automation and amplification of that disastrous unchallenged leadership mode that Lesseps exemplified, which always leads to disastrous consequences.

Lesseps said, like the tech titans, that the Panama Canal would generate wealth and broader benefits for the nation.

Power is the ability of an entity to achieve its OWN ends.

With the continuation of the echo chamber and actions, powerful leaders are obeyed because they have defined and fuelled the norms and beliefs which confer great standing and prestige upon them. It's this very prestige that people turn to and consistently refer to as a defining secular measure of success, which determines who is at the top of the hierarchy.

Charisma is a simple idea, a nice story, and if it has a ring of truth, it is more likely to spread, in particular, if advocated by people with the right social status.

Ilana Redstone
Associate Professor
University of Illinois at Urbana-Champaign
October 2019

Twitter: @irakresh
Web: www.diverseperspectivesconsulting.com

The distinction between perspective and truth has become ever more critical since the level of political polarization and the degree of hostility between political parties has increased. The combination of a critical mass supporting the on-campus worldview and a moral judgment levied on those who deviate have muddied this distinction precisely at the moment when it needs to be made most stridently.

Napoleon was a great orator and persuader. After his attempted arrest in Grenoble, he talked himself out of his arrest and became the ruling prince.

Such lofty influence was driven by the US financial sector, which, in 2000, drove 40% of USA corporate profits; so with it came blind spots. During the financial crisis of 2008, the US financial sector was not just judged to be “too big to jail” but “too big to fail”. This left the bank executives covering their losses, immune from prosecution, even by the US Department of Justice. So generous bailouts were provided to

> If you’ve ever read “What Technology Wants” by Wired founding editor Kevin Kelly,
>
> **When nature causes us a problem, technology is the answer. When technology causes us a problem, technology is the answer**
>
> thedrum.com/opinion/2019/10/25/the-rise-technology-good

cover up the chaos. Little seems to have been learned.

The right ideas with the right ring to them, prevailing beliefs, timing, and attitudes of powerful people and organisations determine which ideas will seem compelling and which can be safely ignored.

Yet, bad but catchy ideas like “more technology will fix it” or “technology is inevitable” create crazy fads amongst investors that pump up the valuations of failing loss-making companies that are fundamentally broken. For example, with Revolut, customer recruitment costs have far exceeded revenue for some time, and the baseless fad of WeWork claiming to be a tech co.

> **Michael Jackson** • 2nd
> Venture Capitalist
> 1d
>
> "the partners recently combed through their companies and identified the 'gross margins' — a measure of profitability — for each one...this was not something the firm regularly looked at."
>
> http://j.mp/33A11E9

There is a natural phenomenon that the rich get richer. Yet, Mai Son seems to have been affected by the same hubris of these scalable, non-compelling SAAS companies that seemingly tend to lose an increasingly large amount of money. This flawed fuelling of these businesses happens because the prevailing mantras and fear of missing out cause blind spots, so the normal investor checks and balances of the business models or their future viability are waived.

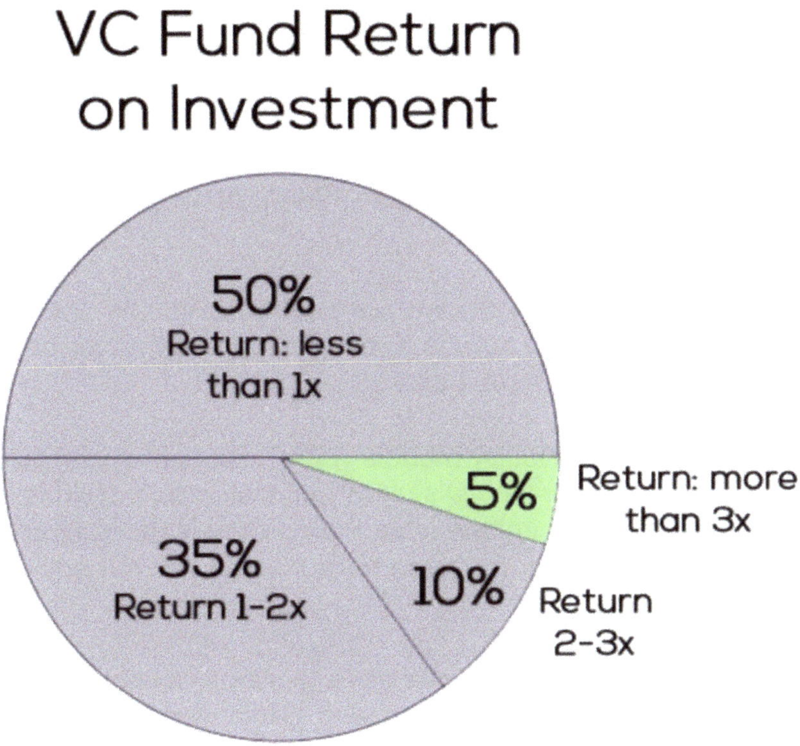

Source: *Techcrunch* and Gil Ben-Artzy, *Money Talks*

So, what are the criteria to become eminent in a technology-driven society?

The more the idea is repeated, the more it is heard by different sources, and the more plausible it becomes. Influence depends upon whether existing organisations and institutions amplify your voice or whether you have the social status or skills to counter or present objections to the prevailing wisdom.

We observe and imitate, and those without a spiritually and morally sound ethical compass are prone to believe anything. What we learn is not good for us but is what others would wish us to believe. People listen to those more eminent in society because they have proven what they are doing works. It's a reasonable secular ethos to follow and aspire to those with more prestige than us because ideas and recommendations from those with status may be worthier of attention.

In a spiritual sense, this logic is turned on its head. Jesus, our creator, said, "the meek shall inherit the earth". So, as a hubris check, do we think we know better than our creator? Where do we find our source of wisdom, purpose, and hope?

Jesus is the original role model we were asked to imitate, yet he was a man without social status, born in a manger, persecuted for his teachings, and died for us on a cross.

But people still imitate secular models. Why? Relative social status, or an adult in the context of a child's view, has status as a child is inclined to suspend disbelief and imitate what their parent is doing because there must be a reason for it. These form the basis of our social cues and who we imitate.

We heed the views of people we believe to be successful, and we judge who is successful by watching who is following who, which follows evolutionary logic. This leaves us vulnerable to absorbing information from those we consider powerful. It has been proven that people still hang on to information that they know to be erroneous from people they imitate.

So, if you have the power and status to set the agenda, you will be listened to. The American bankers who caused the banking crash in 2008 got away with it because the American public put huge value on material wealth. This framing of these bankers with the ultimate status precluded these bankers from going to jail or firing them or even withholding the bonuses of the erroneous bankers, and even these bonuses were funded by government bailouts: AIG received a $180bn bailout in 2008 and yet was allowed to pay out $500m in bonuses to those who wrecked the company.

Infectious and dominant ideas start if you passionately believe in the idea and the idea infers progress. This mantra then defines the way you see reality and filters facts and/or the prevailing dialogue as the idea takes hold and spreads.

The boss of Goldman Sachs believed so much in the banking mantra of wealth, common good and past success that fuelled the banker's untouchable social power that he believed "they were doing God's work". Anyone who questioned that psychology was met with righteous disdain.

Political and economic power drive influence and, therefore, matter because they define who has a voice, sets the agenda, and controls who sits at the table. As Lord Acton said, "Power corrupts, and absolute power corrupts absolutely". From a Christian perspective, this Godless *modus operandi* is the basis of what fuels a vicious cycle of the diabolic.

Dacher Kelter, the social psychologist, concluded, "The more powerful people become, the more selfishly they act and the more they ignore the consequences upon others." They are more likely to cheat or take something, unrightfully claiming that as well as being greedy, they are taking their just desserts and that being greedy is OK.

Keltner observed that people with more expensive cars cut off vehicles and pedestrians 30% of the time, and those with the least expensive cars almost never did.

Paraphrasing Mike Tyson, he said that you "only take offense when you have an inflated view of yourself" or, to put it another way, lose your sense of reality by showing hubris, not the more realistic trait, humility and love for others.

There is a simple solution to this: love God and love others, or else human beings will fill the God-shaped hole with greed, pornography, drinking or sex, etc. The diabolical list of alternatives to fill that gap that God provides is endless.

Valuing the truth is essential for all of us to find humility and the courage to forgive others, or our judgement will become overwhelmed with anger and/or maybe steered by feelings.

Persuasiveness comes in two forms: persuading others and persuading yourself that you are correct. In that way, you create a layer of pride that covers your ideological lies. You will defend these lies at the cost of others' wishes and interests, irrespective of circumstances. This is the basis of tyranny because if the perpetrator has to face their own lies, their scenario falls like a house of cards, so these lies, when challenged, are often masked by outbursts of anger to protect such blind spots and lies.

This is the way that people with the vision dominate the trajectory unchecked.

Relentlessly searching out the truth by knowing how to identify misleading dialogue and challenging flawed dialogue is critical to keep an organisation and dominant individuals in check. This can be done in a safe way for a whole board using the winningthinking® game changer, as I describe in the experiential learning book *You Have Infinite Capability*. But unless you practice different perspectives and realise a different outcome, people won't learn.

These statements imbibed by the less well-off shift their attitudes and intrigue with greed too. The hubris of the elite knows no limits currently by taking the God-like perspective that in some way we can prevail over nature. If these positions are challenged, you will be labelled as a "climate denier", "unimportant", incompetent, or out of touch, etc.

The vision of persuasive people justifies everything, and facts, therefore observed, are used to justify these positions and disregard others. The mantra escalates to prevent thinking and fuel silence as it shifts from 1980s climate change to global warming to climate emergency and now to climate boiling!

Briefly let's take the facts surrounding climate change: the top-level climate strategy that C02 is causing a global emergency is simply false.

The climate shifts through natural cycles, with cooling and warming moving up and down, and this is a normal, natural cycle we are in. So let's get some perspective: C02 is a trace element that only makes up 4% of the 750GT of CO2 that is man-made. The rest comes from natural sources. C02 is a trace element; now 0.04% of the atmosphere is CO2. In the past, that was as high as 0.2% C02. CO2 isn't a crisis. Its main purpose is that it is a natural plant food. The sea level has been 100m higher and 40m lower than it is now. The main issue is protecting nature and economies from sea level rises.

Lewes in the UK was on the sea in Roman times yet is now well inland when the temperature was 5 degrees Celsius higher in Roman times.

So, this lies-based ideology drives simplified solutions and silences dissent – without discussion. The WEF see the value in a crisis to manipulate and push a flawed tech elitist agenda. As Klaus Schwab suggests, "We had Covid, now we have Climate Emergency". He is using both to create the window and justify the acceleration of the Great Reset.

The ideological outcome is a disaster for our economies, pursuing EVS, Hydrogen and Carbon Capture, all of which cause an increase in energy and energy crisis and geopolitical crisis, as well as handing the keys to China. These ideological "solutions" lose sight of any proportionality: take the UK automotive industry, which may reduce 0.01% of LOCAL emissions (forgetting the Chinese emissions making the batteries or the pollution footprint of EVs), costing 800,000 UK jobs in the auto sector. R&D spend matters. It is estimated that 60% of R&D is driven by the automotive industry in Germany.

The IPCC Data is correct, BUT politicians write the conclusions that form the IPCC recommendations. Politicians are being increasingly manipulated by global bodies such as the World Economic Forum, which are experts at persuading people en-masse to believe their flawed ideology to meet their own underlying objectives such as "citizens will own nothing and be happy" and "people therefore over-population are at the root cause of the climate crisis" and "technology will enable" the Great Reset.

These ideologies are so prevalent that the economic interests or wealth of people, which is critical to pay for any such climate improvement, will be dashed by trashing our key industries and handing the keys to these global industries to nefarious parties such as China. China is the expert at persuasion, manipulation, tech control, propaganda and ideology.

So responsible, balanced behaviour doesn't happen naturally. In fact, without an ethical or spiritual perspective of what is good or bad, the globalist strategies take on a draconian and diabolical twist. We cannot expect people who hold great power to adopt social responsibility if they have strong visions and dreams of shaping the future. In the case of Margaret Thatcher, at the end of her tenure, she became autocratic and overzealous. Tony Blair never accepted and still does all he can to justify invading Iraq on the flawed premise that Iraq had biological weapons, yet none were found.

So, the power to persuade corrupts unless you have a strong spiritual compass and/or the leader surrounds oneself with a range of people with different views to break the monopoly over the agenda. The spiritual basis is critical to avoid the diabolically fuelled social norms shifting from what sounds good and true to something truly diabolical, which is what we are witnessing with the "Great Reset" normalised under the narrative of "Climate Emergency "and "gender ideology", cancellation and child mutilation" normalised under the narrative of, equity diversion and inclusion.

Mao, Hitler, and The Climate Change movement all adopted a childlike Greta Thunberg wearing pigtails, and all claimed to be acting for the greater good. Yet, their underlying mechanism is victimhood, scapegoating and terror (blatant violence, and the cruelty of marginalisation and cancel-related terror).

Democratic institutions are critical.

We have witnessed a shift from the Feudalism of the Middle Ages to capitalism and democracy, fostering innovation and economic wealth, and now we shift back to digital feudalism, where the digital platforms such as Amazon own the land and take not 15% of the rent but as intermediates of a sale take 40% of the rent from the occupiers. I call this digital feudalism parasitic tech because it feeds off the profits of the capitalist or productive part of the economy providing the service.

Democracy includes and provides a means for looking after the poor, reducing inequality, and productive innovation creates opportunities. Innovation occurs because ideas take root through freedom for the innovator and market to adopt the myriads of ideas, a few of which thrive and grow into large companies, but most fail.

Democracies constrain oligarchies, inculcate law-abiding behaviour, and create opportunities for all. The test of a democracy or tyranny is whether a single voice dominates, and you can speak up against the prevailing vision without being shot down, or people value others for who they are and take an interest in and value their different views.

In a crisis, a single charismatic voice with a vision can become very attractive, especially in hard times when people are looking for a refreshed purpose. The climate fuelled the rise of the Nazi party and Hitler's dominant vision of the German nation's rise and dominant mantras blaming the Jews for the post-World War 2 failed German banking system, which led to the scapegoating of the Jewish people that ended in such appalling terror and murder.

The jury theorem depends upon a variety of different views to reach a democratic and balanced conclusion. Its success is rooted in accountability and trusting people, or the antithesis of a technocratic dystopia that constitutes social media and imbibes it's necessary cancel culture.

Democracy unwinds itself by, for example, delegating AI responsibility to technocrats or working parties of technocrats that think the same, so the public don't get involved in the details of the government or technological affairs. These groups are very often dominated by Big 4 consultants, which act in their own interests of repeat revenue, fuelling the issues and flawed ideologies. So, by not resolving the issues, the Big 4 create endless rent or repeat business. These groups often get usurped by mantras that suggest “AI is inevitable” and “is good” or “big finance” or “Trust in AI” is “AI is good”. So, the march continues unabated, and any challenging robust analysis is marginalised.

Democracy avoids the focus on these narrow visions, which undermine the inclusive foundation of a sound vision. Still, the technocrat’s mantra is often that ordinary people don’t have the knowledge or vision to have valuable views, so catch-22. This undermines Yanis Varoufakis’ view that an individual solution may be to create shared ownership of the parasitic tech company so that the masses can vote and influence the direction to everyone’s benefit. That is highly unlikely for the elite to welcome if the masses are considered incompetent and lacking the vision to make a useful contribution.

The direction and application of technology determine the outcome and who wins and who loses. What determines who wins and who loses is who has the power to decide, and the tech elite WEF want to retain that power and wealth.

People with access to the corridors of power are increasingly the dominant FANG technology oligarchs and/or entities such as the unelected WTO, WEF, WHO, Blackrock, Vanguard, ESG types, etc. These are funded and fuelled by Oligarchs, consultancies, and others in support of their vested interests and dominant visions that undermine democracy, control the dialogue, and undermine a sound outcome.

The success of the oligarchs is therefore a vicious cycle because the process and autocratic actions fuel additional power and wealth, enabling them to become more powerful.

Because technology can be so flexibly applied, new stories/dominant narratives cover up blatant failures. Take the collapse of crypto, for example. Even after its failure, the dominant underlying narratives, and claims of individual freedom from government control and protection against government-fuelled inflation remained intact, with the crypto advocates claiming “there were just some nefarious players” who caused the failure...

Currently the gains accrued to those controlling the technology are harming and numbing the rest of the population. So, the corrupting dominant narrative must be recognised and addressed, and healthy, varied dialogue, a broad set of ethical interests and critical thinking must begin.

The silencing that misaligns the actions of the dissenters leads to menticide or people not caring and becoming angry, with a trend for people not to believe anything they are told and entrapping them in their own unhealthy, unchecked ideological and atomised bubble. Tragically that leaves them isolated, marginalised, unable to collectively bargain, and very easy to manipulate - the exact opposite of their good intentions but perfect surfs for the AI/technology feudal system!

Serfs in the digital feudal system are unpaid. They are unpaid for creating the data, which forms the basis of the digital capital. In the past, 80-85% of workers' output in the Middle Ages system went to workers. Now, Facebook returns less than 1% to users/serfs for uploading content to maintain the value of the digital asset or "land (*1)".

The equivalent of the land (*1) in the previous feudal system is the digital platform such as Amazon that adds no real value yet takes c40% of the sale.

So, this capital is created with unpaid labour, which shrinks demand, so banks print more money to replenish demand. There is no work and play – no demarcation, so the workers become angst-ridden, looking at social media without autonomy, no social democracy, or the option to opt out unless they can afford to become a Luddite.

A capitalism system is made up of two parts: a market (where products or services are sold or purchased) and profit. In a digital feudal system, both are gone; the digital elite collect rents in a monopolistic way, taking rent for no effort - not profit. This depresses the demand to create digital fiefdoms. EV manufacturers such as Volkswagen have no cloud capital, so they are collapsing. Conversely, Tesla and the Chinese EV OEMs have cloud capital, so democracy will not work.

Yanis Varoufakis argues that the solution is to socialise this digital "cloud capital" to people using shares. Employees can't be traded to give a vote as an owner and input to steer strategy. AI turbocharges tech feudalism. Yanis argues that if we owned a digital identity, we could bid off platforms with our I person vote, and if we had a free digital bank account, we could transform the market and system. He admits that he is addicted to the digital system. I would argue that in his digital bidding and voting space, products will become increasingly commoditised, less local and more transactional - less relationship-centric - so profits will be trashed unless you falsely control this utopian market system.

> This is a long way from true capitalism, as Graeber notes, and actually looks more like classic medieval feudalism. Much within the modern corporation is less about making things or solving problems and more about the political process of gaining control over the flows of resources. The result is a proliferation of jobs that actually serve very little if any economic function, and only make sense from the perspective of rent seeking and power relations. Many like to laugh at the absurd inefficiencies of the Soviet Union, where so many people only pretended to do useful work, yet this may be significantly true in Western economies as well (only in the West they actually get paid for it).

Source Economist

For most people trapped in the UK's Norman feudal system, the adoption of technology deployed in industry deepened their poverty, and lords imposed more onerous labour requirements as well, so malnutrition worsened. In the 1300s, the people were very vulnerable: the Black Death took 30% of these chronically malnourished people's lives.

The UK clergy made up the elite of 5%, yet they captured most of the agricultural surplus, with which they built cathedrals, and the 1300 bishops etc, owned 30% of the agricultural land.

In France, 20% of the output was spent on church construction projects. Even monarchs were not strong enough to take that revenue from the church. They maintained control through violence, which led to the Peasants' Revolt in 1382 when they were no longer willing to be trapped in servitude and became determined not to give way until the magnates of the feudal system were destroyed. They ultimately attacked the Tower of London, where the king was taking refuge. This led to Richard II agreeing to the demands to abolish their serfdom.

The church's persuasive dominant narrative and mantra at the time was that they were God's emissaries, so they could not be questioned or held to account for their crimes. This led to Thomas Becket's killing.

Netflix and Disney Shares Fall After Apple Unveils $4.99 Streaming Service

Apple CEO Tim Cook announces Apple TV+ during a new product event Tuesday, Sept. 10, 2019, in Cupertino, Calif. (AP Photo/Tony Avelar) ASSOCIATED PRESS

The parallel with the current technology system/WEF diatribe of "common people owning nothing and being happy" is that the social power in medieval Europe enabled the elite feudal lords to remain wealthy and trap the poor who were coerced to work unpaid on the lord's property. The Lords controlled prices and banned the independence of home milling, and windmills were demanded to be demolished because they competed with the established watermills.

This is an exact parallel with how technology companies such as Google control the marketing on their digital land/platforms for SME businesses today and how the FANG companies buy up their competitors as did the medieval landlords.

The clergies' rights in the Middle Ages too were decided in a church court, so outcomes were ruled in favour of the abbot, exactly as social media companies judge who should be included in the social process and who should be punished and/or cancelled with impunity.

Removing support for Google Play and Playster

GOOGLE PLAY

Google Play's current terms of service create a relationship with distributors like D2D that is neither manageable nor scalable, forcing authors and aggregators both to take extreme measures to be listed. Pricing changes, requirements for account management, and other restrictions imposed on our authors make Google Play an unattractive option.

Effective October 30th, Draft2Digital will no longer offer Google Play distribution, we will be closing our Google Play Distribution Beta program, and all titles will be taken down at that time.

"Blessed are they which do hunger and thirst after righteousness: for they shall be filled. " Matthew **5:6 KJV**

New technology favoured the elite in medieval feudal times, and productivity gains and benefits were retained for the elite. This was reinforced by corrupt court justice and coercion, an exact metaphor for the technocratic oligarchs and their ideologic rights to define hate speech, truth, etc vs the dominant mantras we have today.

Malthus's narrative was that the poor were responsible for their misery and "the poor were feckless", and if you give them more land for a cow, they will just have more children. This is a metaphor akin to Claus Schwab's evil mantra of "being happy and having nothing" or the Malthus narrative parallel: "as a population increases disproportionately vs subsistence or their ability to provide". Do you see the link here between people and excessive population (of idiots!) that the WEF claim is the root cause of the climate emergency?

In the last throws of the feudal system, labour shortage created collective bargaining power for the peasants, enabling them to refuse to pay fines and walk away to other manors with higher wages. Gower put it as: “servants are now masters and masters are now servants”. So technology that reduces tasks creates a reduction in collective bargaining too, trapped in the hellish technocratic virtual reality of ideology.

Single crop-based agriculture intensified the issue, as do the vast, powerful monopolies that make up the FANG-type digital systems or the dominance of Blackrock type Oligarchs that drive their ideological ESG agendas to maintain the same type of feudal control and pursue their own narrow agendas whilst retaining their immense narrow power.

Artificial intelligence, as we shall explore later, has the capability to replicate and multiply the number of dominant egos online using bots. This thereby manipulates and overwhelms individuals using the oppressive WEF dominant type of corrupting narrative, causing devastating damage to democracy.

People adapted to early technology not because they were controlled daily but because they were coerced in the background with persuasion centre stage. Being controlled hour by hour by social media, ideology, common psychology, dominant narratives, echo chambers and turbo-charged AI provides the ultimate set of tools to addict and control subversively and nefariously at an individual level.

The technologists in the Middle Ages linked and switched their arguments based on their technological interests with what they claimed constituted the common good. It was only when the hold of tyrannies weakened that, in 1600, wages started increasing for the people. Such modernisation was far from equal; expanding commercial farms meant larger farms, which meant taking land from peasants and removing their rights, exactly as the WEF propose by reducing the wealth of the middle classes through their climate agenda.

In the 1600s, only 10% of the male population had a vote, and Malthus indicated that it was more humane not to let the living standards of the poor increase, lest they had more children.

Young, in 1771, claimed that “the poor must be kept poor to remain industrious”. And those who spoke against Young were “contemptible cavaliers”. The self-interest of taking land away from the poor clearly benefitted those doing the taking, and that was a technological choice. Yet Young changed his mind and concluded that improvements could have been gained without driving the peasants into misery.

This is the same logic as the persuasive objectives of the WEF: “You’ll own nothing. And you’ll be happy” — 8 Predictions for the World in 2030, WEF, 2016.

The World Economic Forum (WEF) is an independent international organisation committed to improving the state of the world. As stated in its mission, it *“engages business, political, academic and other leaders of society to shape global, regional and industry agendas.”*

Hammond, the Governor of Carolina, made famous the doctrine of “positive good”, “but slavery is no evil”, claiming that slaves as a class could not be a more contented race on this earth. This narrative is the same type of dismissive WEF dialogue: “You’ll own nothing. And you’ll be happy” — 8 Predictions for the World in 2030, WEF, 2016

The Soviets used coercion to justify creating their ideal type of socialist society, and Trotsky argued that forced collective agriculture was the way forward for the Soviets. This is a very similar approach to Hammond’s on the USA plantations. So, when the kulaks' peasant farmers became wealthy, they were viewed by Stalin as a major anti-Soviet force. The Soviets believed that the US approach of coerced labour and mechanisation was the way forward, so they copied their approach.

But according to an estimate from the 1930s Soviet output, it would have been up to 45% higher without centralised farming. There was a staggering human toll of five million Soviet deaths of a population of 150m in 1933. So, like the European model, there was no improvement for the workers. However, rather than being in the interests of religious elite, technological progress was justified by being in the interests of the proletariat and Communist Party, so once again, it was a socially biased outcome to benefit very few.

There is an interesting paradox here because the woke neo-Marxist iteration and division centred around gender, race and sexual group identity is aimed at creating a divided platform that cannot collectively organise themselves to improve their lot vs the lot of the proletariat. This neo-Marxist iteration replaced the Marxist view related to the now-dated notion that focused on an economic improvement for the poor that the Marxist ideology set out.

The problem with these systems is that they don't allow for discourse, so when talking stops, violence starts, so the Soviets shot dissenters or sent them to Siberia, with a million killed in 1937 and 18m were sent to Gulags between 1930 and 1955. The Soviet propaganda was akin to the current stated "technological improvements" that sprang into action and were marketed as progress. Stalin's propaganda claimed that the system's success relies on the "voluntary character of the collective farm movement, and farms must not be taken by force. That would be foolish". So, in the same way, the elite chose the path for the technology based upon (and justified by) their own vision, millions of ordinary people paid the price.

The current narrative from the current technology elite is that spectacular advances are around the corner, and more technology will fix the issues. But in the same way, there is nothing automatic trickling down to benefit ordinary people because the masses are being manipulated and trapped and atomised in a social media dystopia that makes it almost impossible to organise themselves, rise or challenge the narrative.

And those who take steps to prevail get knocked down very quickly and hard and are cancelled without justice or impunity. As the tech elite's "results" get less apparent, the rate of investment increases and the narratives defining progress and the common good remain unchanged. Those with the insight and courage to speak out get punished, exposed, and proven guilty and cancelled before being proven innocent.

M | **Bloomberg**Opinion

I might be the only person on Earth feeling sorry for the big boys of technology. Jack Dorsey from Twitter, Mark Zuckerberg from Facebook, all those Google nerds: They're monumentally screwed, because they have no idea how to tame the monsters they have created.

Little did they know or care what problems they were creating. And now, they're being asked to solve — or acknowledge, or something — — some really big issues, such as what to do about people who use their platforms to meddle in elections or spread lies, paranoia, bigotry and straight-up hate.

Instead, they're manufacturing baloney explanations about how they'll use more technology, or maybe more people, to handle the civic duties they had hoped to avoid.

The technology advances look after the interests of the powerful elite, those controlling the vision and narrative, so the disenfranchisement of the masses makes sense. As productivity rises, powerful people extract more effort from the unpaid digital serfs and productive industries hand over more of their production.

M | **Bloomberg**Opinion

Algorithms can't comprehend truth. They just repeat the past. If we train them to delete tweets with a specific word in them, they can certainly be automated. But the malevolent actors who want to blow through an automated censoring algorithm will do so easily. Until we have a model of truth that is much better than what we have now, there's simply nothing else to say about it.

Social media is doomed to a tyrannical trajectory because communication without a person-to-person relationship doesn't work. People don't listen to a different view from people they don't have a different relationship with. So, the outcome of access to a mass different view without effective communication has no alternative to flit to violence or the violence of marginalisation and the dehumanisation of cancelling. Denying someone their view is psychologically the inner equivalent of suicide and/or the stress of being forced to comply with the propaganda.

For deviating from the various criteria, hate speech, offence, misinformation, speaking against the prevailing wisdom, vaccines and or "climate emergency", etc, online cancellation is the only way to avoid escalating conflict online by censoring and normalising the outlying views. In parallel, this unregulated cancellation fits well with the tech elite who achieve their unpopular opaque agendas by leveraging their tech control interests to control their agenda and narrative. The tech elite can define what constitutes the party line with dissenters cancelled.

Furthermore, the dissenters are the minority of those who care and have the courage to speak up. These people perform a vital role in society because they relentlessly search out the truth, and if this is denied and they are marginalised and demonetised, they will become angry and subverted. They won't, however, be able to lead the collective bargaining in the way they would have done to organise and leverage an alternative strategy, such as voting rights and joint ownership of tech platforms to steer strategy, as Yanis Varoufakis suggests.

In summary, most of the world has been worse off than their foraging ancestors, so we should carefully examine any ideas that purport to be progress, particularly when powerful groups or persuasive people set the agenda and vision.

Digital technology has undermined sharing mechanisms, and with artificial intelligence, our future is turbocharged, and the influence of the tech elite is multiplied. This takes us towards our feudal agricultural past of economic rent and feudalism ie, a parasitic digital feudal system that feeds off the productive parts of the economy with the value of the digital platforms (land) being maintained by contributing to its value through social media, by the unpaid surfs.

Financial Times

He quickly realised that the batch of coffee he had just sold — which was roasted in the US — was generating about $30,000 in retail sales of which he received just $600.

Mr Lander says that like Thrive, many coffee companies from large to small have their own programmes to help the grower, but acknowledges that something more structural across the industry needs to be put in place.

Nadella said in his book *Hit Refresh* of 2017, “we hope to deliver a highly effective alter ego, an agent that knows you deeply”. Currently, Microsoft are releasing Copilot, which does exactly that.

These Copilots can discover scary outputs, such as using confidential information on people’s computers and personal details. The level of Microsoft’s domination of the AI stack-fuelled business systems that operate alongside business users will probably lead to them abusing their market power by integrating these services alongside their 365 suites of products.

Copilot creates proposals and pulls in data from spreadsheets to form complete documents. So AI use alongside leaders has a monumental impact that will shift the focus from people creating strategy to people checking the AI-driven strategy/outputs from their “Copilots”.

Material that is created by AI is not unique or incrementally value-creating. Most of these models are trained in copyrighted material, which users may inadvertently reproduce. GitHub employees were sued for breach of copyright, so Microsoft has agreed to cover legal costs related to Copilots if they use Microsoft’s tools, such as filters, etc.

Shoplifting is described as an epidemic. The BRC data shows an increase in the UK of 24% in the year ending March 2023. There too is a reduction of customer civility shown to company representatives. Shoplifters are repeat offenders, often with drug and alcohol problems: the diseases of despair that have been marginalised in part by technology and less opportunity from society. People despair due to the lack of human contact with organisations’ customer service and/or despair is clearly induced by dealing with the inhuman, fake human replica, uncaring automated systems.

Ideology is escalating tyranny and keeping the best people out of an organisation’s decision-making too. The Heritage Foundation found that out of 65 universities, for every 100 pupils, the universities employed 3 to 4 DEI staff. Many of these 65 universities insisted that candidates submit statements describing their commitment to EDI.

This is like China's state Marxist ideology. The treatment of people who are cancelled on social media, such as when applying to return to LinkedIn, are asked in the same way to sign up for ambiguous ideological policy statements as regards hate speech, etc. As discussed later, when we explore potential solutions, clear language is key, ambiguity and generalisations are dangerous, and it is just the same way the Chinese tyranny uses ambiguity and ideology to increase the threat to individual freedom and rights.

AI and technology are creating impossible problems to fix. China would far rather win a fight using propaganda to destroy will and obfuscate the truth than enter a war. Take Taiwan during the run-up to their current elections. Taiwan are being hit by an onslaught of Chinese propaganda claiming America has asked Taiwan to set up biological weapons labs, Taiwan was going to take samples of Taiwanese blood and hand them to the Americans to develop a virus to kill Chinese people, and a Chinese video maker created a video showing an American lawmaker discussing the possibility that America was planning to blow up the TSMC semi-conductor plant. These false narratives portray America, not China, as the island's threat.

This is becoming a major worry for the Taiwan Government who are being asked whether Taiwan should remain aligned with America. This disinformation has severely infiltrated Taiwan's society, and as its multi-channel deception is supported by AI, it's very difficult to trace. It will only take 7% of voters to change the Taiwanese election results. China is piggybacking on Taiwan's fears of being abandoned by losing America's recognition as in the 1970s.

Taiwan wants to avoid accusations of limiting free speech by silencing Chinese media, such as CTI news, which has moved online. This problem will only accelerate with the use of AI. China's insidious efforts to mislead are increasing as things get tougher for China. Chinese exports are falling, and Chinese citizens are hoarding cash because they are losing confidence in the property-based economy Xi has created.

Effective global communication is key to creating the right environment to develop these global technological strategies. China's Chairman of Nomura was prevented from leaving China. China is becoming increasingly isolated because of a huge fall in business travel and tourism. Chinese inward investment is down 90%. Only 68% of American companies in China were profitable last year. With the opening of China's economy, the bounce back has not happened. Chinese business sentiment has continued to deteriorate, and the window of relationship threshold for meaningful re-engagement with the West is closing.

At a local micro level, communication by being stuck with a chatbot is a significant source of customer aggravation. In businesses' attempts to do away with human interaction, 17% of customers admitted to being uncivil with their interactions with businesses. Whether there is a substitute for humans in business communication is looking increasingly unlikely because people seem to be getting more fraught with everyday interactions and the tech-influenced environment we live in. Claims are once again of a 14% productivity improvement, measured by the number of queries per hour resolved. Does this outweigh the downside of the infuriating experience of communication with an AI-driven interface?

China's position as a sovereign debt creditor has increased its sway over the country and its technology decisions. This makes restructuring debt in middle-income countries with Chinese sovereign debt, such as Sri Lanka, impossible. China refuses to sit on committees with Sri Lankan investors, so Sri Lanka is stuck with their investment partner, which puts them at threat of having to follow China's whims, with changes implemented at a time of China's choosing.

Between 1820 and 1930, output per person in Britain almost doubled, making it the most productive economy in the world. These people were not born into the same land-based riches that dominated the wealth of the past. They were a new breed of entrepreneur inventors.

George Stephenson was a new breed. From the 1750s, people from modest backgrounds could dream big and rise fast in Britain. Their ambitions were also centred around technology and solving practical problems, and British society allowed them to realise their dreams. The contribution of the Royal Society and the scientific establishments was zero; they were practical inventors. China at this time was far ahead of Europe in 1500 and led as late as 1700. In fact, for most of the Middle Ages, Industrial Revolution innovations can be traced directly or indirectly back to China.

Why was Britain first? Isaac Newton had to step carefully when it came to religion, and in the textile industry, hundreds of people made contributions to the development of modern techniques. These ventures were financed mainly from retained earnings, alongside loans from family and friends or people in their business community. This social change created a whole nation of upstarts who innovated to develop technologies, so the middle classes pierced through the rigid, established feudal land-based wealth hierarchy.

Most of these were extremely ambitious men who had no formal education, but they did not start at the very bottom of society. They had practical skills and experience in small-scale enterprises before creating larger businesses. Most importantly, they believed that technology could be an engine of progress and social elevation. These people convinced the aristocracy that they should and could be accommodated.

The centralised feudal system created the class system using a great deal of the king's power, whose priority was acquiring territory. These commercial endeavours were not seen as a priority, and they went unchecked. The Magna Carta paid lip service to the people, and parliament granted powers to nobles and the church. Henry VIII dissolved the monasteries and seized much of their land. 2% of the population owned 25% of the land.

James I's mantra was that he had the divine right of kings representing God on earth. This did not sit well with landowners, and it sparked the Civil War in 1642 that sought to limit Royal power. The Glorious Revolution of 1688 followed and centred on the rights of people with property as the basis of political life.

These feudal leaders could have innovated, but that was rare. So, this new group of rebel industrialists led innovation, resourcefully rising whilst the feudal power system declined. Yet the system was still fixated on the idea that the rich with property should be taken seriously. The new wealth-building factories and industries created a modified version of the existing system rather than overthrowing the existing land-centric social system.

The dialogue of the rich classes continued to see the poor as a burden on UK society, as many still do today. This psychology allowed the new breed of entrepreneurs to focus on accumulating wealth without worrying about their employees or communities or sharing benefits until the workers accumulated enough power to change their lot.

The reality was harsh. In the 1800s, 200,000 people worked in coal mining, of which 30% were children. Conditions worsened as pits became deeper, shortening the lives of the workers. When new opportunities arose for these workers, they were given bargaining power and wages could not be kept low anymore. In the 1800s, people with bargaining power could organise and fight. In our tech dystopia now, those who rise are increasingly atomised in their technological bubbles, psychologically manipulated into submission, forced into following the prevailing wisdom or cancelled. They become trapped, marginalised, and stripped of their individual bargaining power. It is those strong individuals who value the truth that led to such dissent.

During the Industrial Revolution, annual working hours increased from 2700 – 3400 hours, 65 hours per week, without a corresponding increase in income, because that was the only way to survive in the new industrial economy.

Early factories significantly reduced worker autonomy and production at home, with factory owners focused on sweating their assets.

Because workers lacked organisation and political power, and employers were powerful, productivity gains were not shared with workers, yet profits were higher. This is an exact parallel with the modern Tech Titans, albeit through automation, they are creating increasingly less paid work, therefore less leverage for workers, whilst entrapping unpaid digital surfs to maintain their platform content.

Arkwright became one of the richest men in England at the time, and although cotton prices fell, there was virtually no productivity improvement. The Luddites were one organised group that challenged the dominant wisdom that the rich should prevail regardless of the cost to workers.

Automation does not increase a worker's value nor provide additional work, so the consequences then and probably more permanently now fuel big winners. This negated the productivity mantra or positive consequences of employees with increasing productivity. So, automation just destroys the overall value for people as well as its distribution.

The same flawed productivity mantra continues in the UK in 2023 and still extols the same type of disconnected link between automation and improving wages:

"We've protected the most vulnerable and lowest paid with cost-of-living support worth £26 billion. Stabilised the economy and people's mortgage rates. A better future is one where our economy is growing faster so that everybody, everywhere, has new opportunities for better-paying, good jobs. The most powerful plan for sustainable growth is to position the UK to fully benefit from the incredible scientific and technological change the world is seeing.

That's why we're:

Increasing public funding in R&D to £20bn to enhance our world-leading strengths in AI, life sciences, quantum computing, financial services, and green technology.
Seizing the opportunities of Brexit to ensure our regulatory system is agile and supports innovation.
Making sure entrepreneurial and fast-growing companies get the finance they need to expand.
Spreading a culture of creative thinking and doing things differently across every part of the UK.

If we're going to deliver this better future, people will have to work hard. I believe good, well-paid jobs are about more than just financial security. They give people purpose, confidence, dignity – the chance to build a better life for themselves. But I also believe that if you work hard and play by the rules - you should be rewarded. Which is why as soon as we can, the Government will reduce the burden of taxation on working people. And it is staggering that at a time when businesses are crying out for workers, and when so many people are working so hard, a quarter of our labour market is inactive. So, our growth plan will look again at how we can support those who can, to move back into work – including through the welfare system. All this innovation will make this country a beacon of science, technology, and enterprise and lift our productivity, raise our growth rate, create new jobs for decades to come.

Prime Minister Rishi Sunak 2023 Source: https://www.gov.uk/government/news/pms-plan-to-build-an-innovative-economy

So, the Luddites were right to be concerned that knitting frames would decimate opportunities for workers, although some new industries at that time were created.

Distressed workers who fell into poverty and debt were forced to work in workhouses to receive aid. These workhouses were designed to be less hospitable than working to encourage workers to go back to work. The workhouses were really a prison system to punish poverty.

Today, we see a raft of careers at threat from automation. 30% of the total of all workers are drivers, all of whom are at threat due to autonomous vehicles. The vast simplified electric vehicles and less local value add vs the Chinese-dominated battery manufacturing (40% of the value add) leave local economies with increasingly dumb, simple jobs that could easily be automated. These roles form the basis of the current wealth distribution and value add of our economies. Many of the highly paid tech jobs will not be open to the masses because IQ will be a limiting factor. IQ is a cruel factor because you cannot improve your IQ. So, a vast array of workers with lower IQs will be permanently marginalised from the working population. The IQ threshold requirement of work is likely to increase, too. This contrasts with Rishi Sunak's tech and economic perceived wisdom above.

Pollution was out of control in the early phases of the Industrial Revolution. The same is true on a far larger scale now with the rise of EVs. China is creating vast quantities of coal-fired electricity to process lithium, a process that consumes vast amounts of water, and the hazards of mining the rare earth metals cobalt and nickel at a reasonable price force those mining industries into less ethical and unregulated parts of the world. The carbon footprint of manufacturing an EV is a vast percentage of the total footprint during the lifespan of an internal combustion vehicle.

The underlying issue motivating the EV industry is far more WEF-centric: about wrestling control and wealth away from the middle and lower classes for China and other players to permanently dominate the value add and retain the value in China, disregarding the indigenous economies or workers.

The Industrial Revolution became a public health disaster. Yet efficiency and preserving the wealth of the elite remained the prevailing argument, under the auspices and flawed psychology that "increasing productivity will increase wealth". The same is true now, regardless of the losses.

The link between increasing productivity and wages is too simple because it assumes that the pricing and demand and economic wealth are maintained. But tech industries/AI etc are parasitic and rent-taking, and just feed off the productive industries and their profits and add no value or new industries or opportunities, so markets are reducing, and investment is diverted from the productive capitalist profit-generating parts of the economy.

Anyone who then or now questions "this progress" and its inevitability was and is regarded as a fool, conspiracy theorist or worse.

In 1850-1900, productivity increased from 90-123 percent. The first five years of "digital transformation" in the UK Government *reduced* productivity by 6%.

So, the productivity bandwagon needs two preconditions to translate to worker benefits: worker productivity gain and labour leverage. If automation predominates, workers are displaced from existing work, insufficient new tasks are being created and overall markets reduced.

Steam-powered locomotives and agricultural machinery created new industries and demand for high-quality iron and other raw materials. Current parasitic technologies and digital industries such as EVs simplify industries and processes, thereby reducing value add and, as importantly, differential value add. This reduction in differentiation reduced the critical value add to maintain sufficient pricing to sustain these industries or reach a viable return on the vast investments. This will drive further automation in these increasingly simplified industries, so it's a race to the bottom for economies, falsely and disastrously justified by "saving the planet". This type of low value add EV industry is a profitless growth model, which suits China but doesn't meet the investors in the developed world investment criteria. Western investors are left investing in the parasitic "pump and dump" Software as a Service (SAAS) type businesses that scale with no transactional cost, regardless of inflation. But SAAS-type businesses fuel the decline in overall value add, or a lack of a compelling proposition or any value add for workers. These SAAS industries are driven by controlling attention to retain and addict users, so they are designed to be addictive and control and manipulate the helpless, malleable users, immobilising the very same workers that should be revolting.

In 1871, trade unions became legal. In 1838, Chartism enabled all men over 21 to vote, from 18% previously, alongside the removal of the property element required to become a member of the Houses of Parliament. JR Stephens, in 1839, set out that everyone had a right to have a decent standard of living. The Chartists were rejected and imprisoned for that view.

Indian railways were built in India not to benefit the local people but to move British troops around to maintain power by responding to local trouble. This is akin to the UK Government leaving the tech companies to define misinformation and what constitutes hate speech, etc, what is allowed and what is not.

The Indian railway network reduced the cost of oppression, enabling a few thousand troops to rule over 300m Indian people. The omission was tragic in 1870, and in Bengal in the 1940s, when famine in India struck, the railway was not used to bring food in by rail, leaving millions of Indians dead.

In other words, the railway technology introduced to India was not used to improve the lot of the people.

But factories made it easier to organise workers, and more democracy helped with sharing gains.

By quashing debate, silence fuels evil or nefarious interests and misaligns strategy at a vast cost to our economies too.

Currently, people are unknowingly falling into menticide because their actions and their beliefs have become misaligned by being persuaded and manipulated without thought, contrary to any rational logic to follow the perceived wisdom relating to an endless list of flawed ideologies such as climate emergency, the green economy, the roll-out of ULEZ, LGBTQ gender ideology, COVID mandates, CRT and the validity of vaccines etc. People are now bent out of shape and losing themselves, their integrity leaving them silent, pathetically vulnerable, malleable and fearful (so not thinking), so they are like rabbits caught in the headlights and not protesting as they should.

Skills training in manufacturing created opportunities for growth for all uneducated people with the latent capability and headroom that could be unlocked. Henry Ford leveraged training to create new value add, increase productivity, and create new roles. Productivity creates purchasing power too, reducing costs.

The hard times of the 1930s, as now, fuels nationalism. The appeal of charismatic persuasive leaders and extremist power takes the population's focus off their own impossible plights by providing a shiny new energising purpose and refocusing on renewing people's identities. Conservative governments with sensible agendas often lack that compelling appeal.

A disparity in wealth always leads to tyranny to maintain the power of the elite and keep the masses down.

In the 1920s, Sweden recognised the rise of extremism, so it moved away from Marxism to form a coalition of workers. They focused on higher wages and profit sharing, industry wage setting, and social insurance programs. Hansson stated that the party would not aim to support one group at the expense of others. The top 1% of Swedes went from owning 30% of the wealth to only 10% in the 1960s.

People from the Great Depression were so jaded from the experience that they became unwilling to take business risks. By the 1960s, with auto production at 8 million per year, the automobile industry made America what it is today. Wages rose 3% per year from 1950-1970 because the *direction* of technology created new jobs and tasks for *all* skill levels. There was a framework to share the profits with employees and managers. Clerical and professional occupations materialised, most of which did not exist in the 1940s. This demanded more literacy, numeracy, and social skills.

Through AI and digital feudalism, this trend is automating such processes and roles and reducing these opportunities. Post-war was a philanthropic time when the unions allowed automation by making sure it brought benefits and new jobs for skilled and unskilled (lower IQ) workers. So, the decades that followed were unique in creating

Figure 8: Given automation, what are the characteristics and skills believed to be most needed for the future workforce?

(% of respondents)

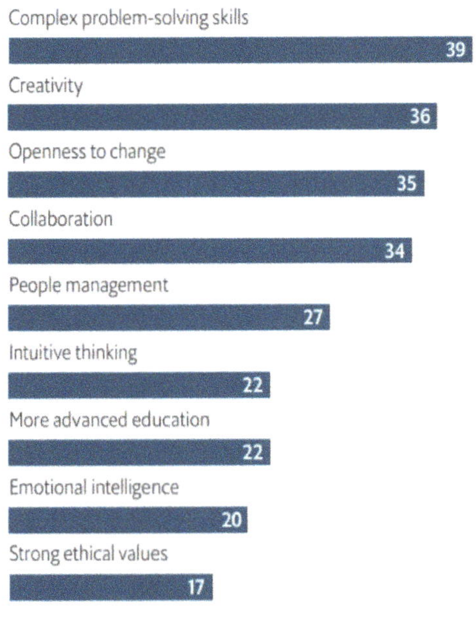

Source: The Economist Intelligence Unit

rapid *shared* prosperity. Albeit, at that time, women and black Americans did not share the wealth equally.

Digitisation and rent-sharing

Technology and computing extrapolate the perceived wisdom because the wisdom is the choice the technologists have made, and the wisdom is the basis of the technology input.

1960s Silicon Valley's focus was decentralisation and freedom as it is today. The same weird freedom-centric psychology from government-induced inflation and anti-corporate bureaucracy thinking is behind crypto and Blockchain, so it lives on today! Big technology companies like IBM became bureaucratic. Secrecy was viewed as the keystone of tyranny, so the focus was to democratise information and break the grip of IBM, etc.

In the event, digital became the graveyard of shared prosperity. It automated the work of the most disadvantaged labour and favoured postgraduates. The psychology that fuelled this was like feudal times, maximising profits and shareholder values for the "common good", the same organising psychology principle and blind spot as it is today for much of society.

This was very different from the libertarian utopia. The new vision from Silicon Valley was a digital utopia based upon top-down software to automate and control labour. US wages grew from 2.5% per year to 0.45% from 1980, with productivity growth at 1.5% from 1980 to the present. Wages for graduates rose and workers fell 0.5% from 1980-2017. And the wealth of the top 1% grew from 10% in 1980 to 19% in 2019. This leaves most young Americans likely to be worse off than their parents.

In terms of rent sharing, the focus on automation and outsourcing shifted the share of national income for workers from about 20% in 2019 to about 10%, and unions are weaker. Germany was at 70% in the 1980s and is now at c60%. In manufacturing, the labour content has dropped by c20% since the 1980s.

With few countervailing forces in opposition, organisations have now turned their back on creating new tasks and opportunities, especially for those without a degree.

Technology is anti-labour, not inevitable.

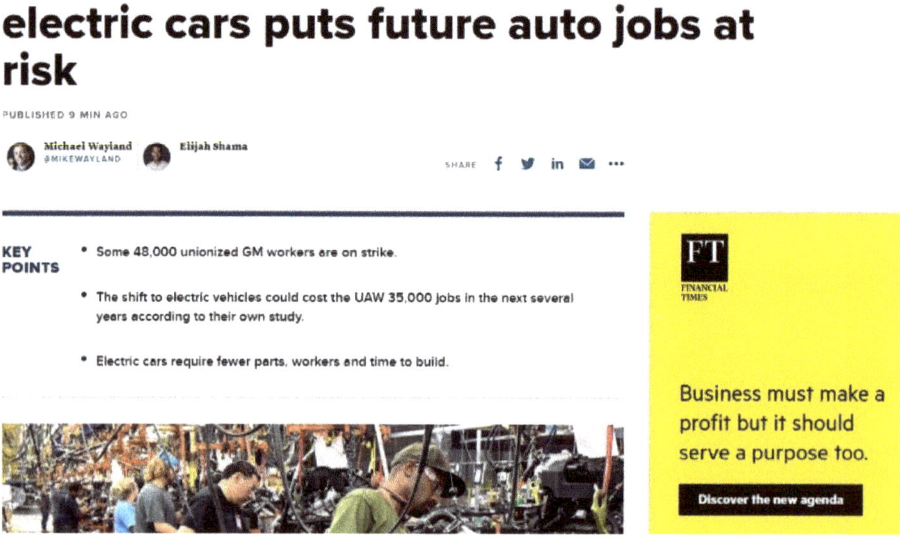

electric cars puts future auto jobs at risk

PUBLISHED 9 MIN AGO

Michael Wayland @MIKEWAYLAND
Elijah Shama

SHARE

KEY POINTS

- Some 48,000 unionized GM workers are on strike.
- The shift to electric vehicles could cost the UAW 35,000 jobs in the next several years according to their own study.
- Electric cars require fewer parts, workers and time to build.

FT FINANCIAL TIMES

Business must make a profit but it should serve a purpose too.

Discover the new agenda

Automation boosts inequality because it automates the simpler work or work carried out by low-skilled workers. Automation is now spreading to clerical and higher-value professionals such as law and medical roles. Offshoring simple low value add products to China costs about 3 million US roles. The core US automotive manufacturing base is now in real danger of being completely usurped by China by handing their control of the core of the auto battery manufacture EV design to China alongside China's vice-like grip on manufacturing because they control the critical raw materials of lithium and cobalt, as well as the 95% of the rare earth metals that form the basis of electronics manufacturing.

As manufacturing jobs have disappeared, the entry-level roles into a career for many less educated people create social problems. Mental health and “deaths of despair” due to drugs, alcohol and suicide are now causing US life expectancy to fall several years in a row.

> Being overly connected can cause **psychological** issues such as distraction, narcissism, expectation of instant gratification, and even **depression**. Beside **affecting** users' **mental health**, use of **technology** can also have **negative** repercussions on physical **health** causing vision **problems**, hearing loss, and neck strain.
>
> Health and Technology — Digital Responsibility
> www.digitalresponsibility.org › health-and-technology

Technology is not inevitable; it’s a simple choice. Imposing minimum wages without profit improvement just creates job losses.

It was institutional changes that turned anti-labour. The Silicon Valley freedom fighters of the ‘60s turned into an agenda for automation and worker disempowerment. Moves to benefit workers and consumers were resisted by corporations organising against liberal regulations, leaving them free to vote for their own business interests.

It has become the norm for government to become intertwined with corporations. Charles Wilson, in 1953, was President of GM and Secretary of Defence, “what was good for our country is also good for GM”.

By the 1980s, this view of corporate synergy with government became commonplace. Favouring companies that boosted profits was the best way to help everyone. So, America rejected conservatism and entered a socialist experiment. They stated the free enterprise system was seen as selfish and materialistic.

The Reagan policies led to the “trickle-down economics” of the 1980s by cutting taxes on the rich, alongside the paradigm of “what was good for business was good for everyone”.

The “productivity” bandwagon was still part of this narrative by inferring “increasing demand for workers and shared prosperity”. This conflicts with consumer activists like Nader and Adam Smith, whose view was if the market worked well and there was enough competition, regulation was unnecessary.

Keynes argued that markets do not work in an idealised way, for example, the productivity mantra collapses when further growth isn't there or when there isn't enough competition for labour.

So, if the labour market turns soft because of ever-increasing automation, wealth distribution will be hindered because the labour force will have no clout. Consumers, too, have a tough time differentiating product quality and thereby value. Hayek quoted the collapse of the central market control of the Soviet Union, and decentralised systems helped share information better.

So, the idea that unregulated markets work for the "common good" became the new basis of public policy. Friedman argued that social responsibility wasn't the role of business. Businesses should only care about making profits as the basis for returns for shareholders. So big businesses were now heroes, not villains, with a clear mandate to increase profits regardless of the social impact. Jack Welch heeded the advice and took a tough stance on wages.

This obsession, alongside the widespread use of executive stock options, fuelled the misreporting fraud at Enron. The risk here is that powerful (flawed) ideology creates blind spots by creating topics or ideals that become unquestionable, and the narrative is forced to fit, hiding and blinding the organisation to their dominant falsehoods.

The Friedman ideals were pushed in business school teaching. By 2020, 43% of CEOs had a business degree, embracing his myopic ideals, such as cutting wages to increase profits.

The oil price shock of 1973 was interpreted as the US economy was broken. This reinforced the Friedman doctrine, so rejecting regulations and rejecting collective bargaining was the answer.

Large businesses control prices through scale and buy out of blocks of competition, so even small groups can do a lot of market damage. Competition regulation was largely deployed to counteract corporations from unfairly increasing prices, whereas there is a more nefarious market threat coming from China and the CCP, where heavily subsidised Chinese businesses are entering markets with a strategy of profitless growth. Seen alongside China's control of key materials, this makes China impossible to compete with, so it trashes competition. This is seen in EV OEMs, bike parts, car parts and musical instruments. These Chinese players are moving upmarket whilst keeping prices low to keep out competition.

A US court justice stated, “We may have democracy, or we may have wealth concentrated in the hands of a few, but we can’t have both.” As corporations become very large, they exercise disproportionate political power. Their extreme wealth created for the owners degrades the political process, which makes it hard, if not impossible, even for governments to counter their vision.

So now Google, Facebook, Apple, Amazon, and Microsoft are all parasitic tech companies, jointly worth about 20% of the US GDP. The market power of Blackstone Vanguard and State Street, who own 88% of the S&P 500 market companies, has an unhealthy monopolistic and ideological influence too. A new US anti-trust approach has taken a more hands-off approach to mergers and acquisitions, allowing more tech giants to acquire companies that could have become their rivals.

Gargantuan monopolies are bad for markets, democracy, social well-being, labour, fair competition pricing and innovation. Tech monopolies, turbo-charged by AI, multiply the wealth for their owners as they grow. When workers organise and protest, they increase the likelihood that plant will close and shift production to non-unionised plants.

CONCENTRATION OF AI ACQUISITIONS BY CATEGORY

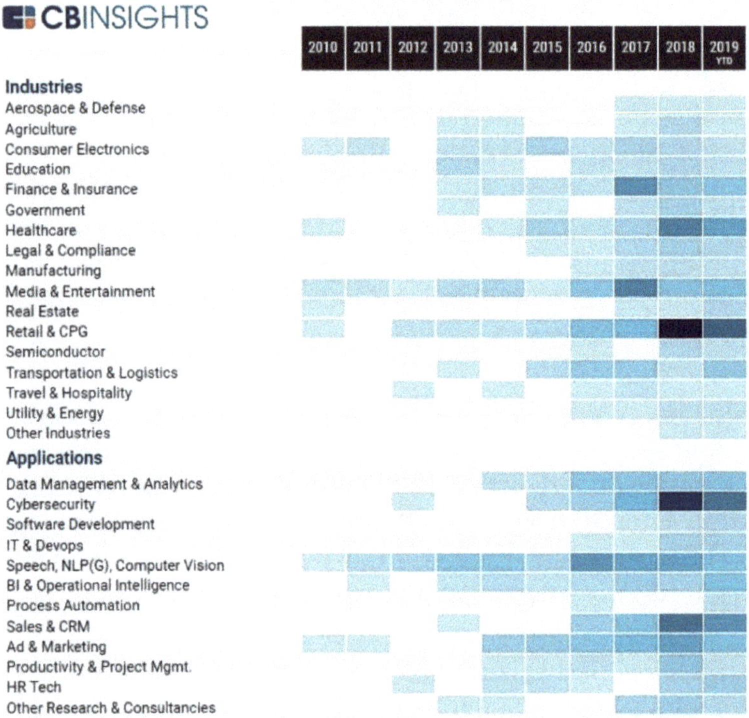

Germany has a long history of works councils where workers are represented on boards. The US banned secondary action, and strikes could only take place at business unit level without industry coordination. Regan described striking workers as a peril to US national safety. In 2021, only 10% of US workers were union members.

Digital technology sits well alongside Friedman's view of reducing labour costs and not a resource. Indeed, the focus on labour cost reduction is now centred on reducing middle management (historically the main protagonists that steered strategy) as well as eliminating unskilled tasks. This led to more complex jobs for smart people. This is progressively raising the IQ threshold for workers, who almost always need a degree. IQ is a fundamental personal characteristic, so it cannot be moved or improved through training. This trend locks people permanently out of the workplace.

Management consultants fuelled this populist anti-labour focus with their myopic focus on making better use of digital technology and pushing the likes of the corrupting and conflicting and silencing equity diversity ideology. These ideas are prevalent and communicated by articulate management experts, and it has become harder for workers to resist. In 1970, 33% of workers were in clerical roles with decent salaries. Now that is down to 19%.

Lee Felsenstein quoted the design motto “designed by geniuses for use by idiots” or the unwashed must keep their hands off rather than allowing the users to learn about and gain some control over the technological tools.

In the 1990s, the US wasn’t under so much labour pressure, so it lagged the automation binge vs Germany and Japan.

Most computing efforts are still focused on automation. German works councils are more focused on redeploying their heavily trained workforce into new roles where labour relations are stronger.

The US tax system always favoured capital expenditure at 5% vs investment in people c15%. This incentivises automation. Reducing government investment in technology development leaves the corporations to determine its trajectory and applications.

Bill Gates typified following the money and social power and summed up the techno-optimism as “show me a problem and I’ll look for technology to fix it”. Whether the technology may be positively or negatively socially biased seems to have no consequence.

The perception of the authoritarian psychology of big business seems to be usurped by “disruption”. Disruption means that disrupting existing practices is good and to be encouraged. Zuckerberg typified this with his mantra, “Move fast and break things”.

Paul Graham from Silicon Valley stated, “I’ve become an expert on how to increase economic inequality”.

The tech elite and leaders still regard that most people aren't smart enough to do their jobs, so idiot-proofing their jobs was justifiable. Little did the purveyors of this mantra realise that treating people like "idiots under surveillance" creates unpredictable results, trashes well-being gaps and escalates risk, so fuels more automation and the self-fuelling mantra that "people are idiots!".

Digital utopia was based upon productivity, and AI upon a customer convenience mantra, under the umbrella that productivity improvement (that has rarely and may never materialise) was a common good.

During the last decades, there have been reduced productivity gains and reduced margins that create wealth to share and very small gains, if any, from digital investments. We may not be appreciating the gains from smartphones and better search engines, but their addictive nature, which is designed to control our attention, probably far outweighs any benefits.

Feudalism – control of our attention

" From the statistics alone, it's clear that social media has become an integral (and to a large extent, unavoidable) part of our lives "

" The desire for a 'hit' of dopamine, coupled with a failure to gain instant gratification, may prompt users to perpetually refresh their social media feeds. "

" Fear of Missing Out has been linked to intensive social media use and is associated with lower mood and life satisfaction "

Tyler Cowen suggests that disappointing productivity growth is due to dwindling low-hanging fruit and an overbearing tech infrastructure wasting swathes of time. The great innovations that have given rise to 2.5% productivity growth vs c0.5% now and slower suggest the mantra of productivity growth unattainable. As most new technology is parasitic and adds no real value on the ground, it is a tech rent, not profit, so it does not realise useful profit improvement anyway.

Investment is unbalanced, with manufacturing sectors lagging. Productivity gains may be limited to new products. The overall value add to an economy, or defendable margin comes into question with the shift to oversimplified digital products because the market is dominated by extreme price pressures alongside limited sources of differentiation, causing an economic low-value add race to the bottom. This leaves the rent-taking, non-value-adding tech companies winning over the productive sector.

Toyota and Tesla reinstated their workers' central roles by investing in their training because they noticed that productivity was plateauing. They lost the flexibility without people in the loop to innovate to be able to switch to respond to changes in demand. Musk, following his automation binge, stated that “humans are underrated”.

Capek stated that “only years of practice will teach you the mysteries and bold certainty of a real gardener, who treads at random, yet tramples on nothing”. Ad how fulfilling this is for people.

It follows that better financial; innovative and productivity results would follow when people became central to processes again. The people and relationships aspects of a purchase drive the customer experience and, therefore, pricing by moving away from selling a widget and from buying a commodity.

The tech mantras determine that these digital solutions, almost without exception, were defined as being in the public interest. But because the mantra creates a blind spot and deceives the audience of reality, the outcome is a dystopian, lower value add, and lower margin transaction and causes people to lose their jobs and livelihoods. The perfect deception.

So, attempts of small tech companies to decentralise the control away from the IBMs and FANGs failed. The best ones that presented a threat simply got pushed out or acquired by the FANG-type corporation.

Then there was artificial intelligence!

It's easy to believe that hindsight is not what it was. The Big 4 and Bill Gates and others argue that the concerns about AI are exaggerated. Gates' system is called "creative capitalism". He set the philanthropic goal for everyone to take on a challenge to alleviate the world's problems. This mantra is an AI illusion that enriches the elite and steers towards automation, surveillance or an environment that kills people's souls. People need space, opportunity, and support as well as being valued for who they are and their differences.

How useful are these tech tools in enabling people to achieve their objectives? The obsession with artificial intelligence creates mass-scale data collection, confuses accountability, and disempowers workers. The mad scramble to automate work is accelerated, even though this is like-for-like automation, so it does not realise productivity benefits, just replaces the person. Large-scale data collection and automation simply enrich those who control these technologies. It is well-proven that people can make good decisions with limited data. Beyond seven data points make no difference to the accuracy of human decision-making. To be automated by traditional software, the task must be routine, but only a small fraction of human tasks *or values* are routine.

People rely heavily on social interaction. Customers value customer interaction rather than rigid "efficient" processes. Flexible, enabled people create excellent customer experience because, given the opportunity, people are a creative species. Customer service requires social as well as problem-solving skills for problems that are usually idiosyncratic.

Artificial Intelligence's ambition is to expand automation to non-routine tasks and identify and investigate transactions that look suspicious. Bots are competing more with doctors' and lawyers' tasks, so the threat is moving up the corporate ladder. FANG companies' CEOs all claim that AI will beneficially transform technology over the next decades. But it's more likely that there will be benefits and downsides and further shifts away from the value of workers, reduced wages, reduced work, and less democracy. This undermines the economic value on which to build our economies.

The AI mantras that reflect a dismissive view of human capabilities wishing people to be seen as error-prone, are made without any idea of what we are as human beings or what drives the best human performance.

The Netscape founder said that 5 great programmers outperform 1,000 mediocre programmers. This justifies using exceptional talent to create top-down systems to control and limit mistakes and the cost of people at work. Then it's acceptable to collect masses of data and replace people with machines, despite this approach only reaching human parity, rather than complementing humans and therefore focussed on cutting labour out.

Hunter-gatherers have a remarkable degree of adaptation to changed conditions. In September 2009, Portishead turned off the traffic lights at a busy intersection. People adapted and used more common sense. At the end of four weeks, traffic flow increased significantly with no increase in accidents or injuries.

Humans solve problems using situation and social skills in their community by taking a view of others' mental states and the context, social communication, exploring the hypotheses that underpin the validity of our decision-making, and sharing objectives through the empathy they have for others' interests. If the value of these social elements is removed through echo chambers, atomisation, or cancellation on social media, this warrants the validity of the human judgement and decision disastrously skewed, regardless of the contributor's IQ.

Social skills are critical alongside human IQ to be effective. People who become atomised are removed from this social safety check and balance or safety net and are very easy to manipulate to believe a dominant or persuasive narrative, regardless of whether it is true or fits their beliefs.

Open communication between employees and this social environment is critical in a customer service environment to improve service, create effective solutions and reduce mistakes. Companies rolling out AI seek people with these social skills. The fundamentals of data-driven AI problem-solving and people-centric, social and community-centric solving are like oil and water. One runs on vast data regardless of the human checks and balances, and the other people run on 5-7 and the power of a social network to solve problems.

AI will reduce the degree of people-centric social interaction, reducing social skills as time progresses and increasing the risk of tech-driven misaligned solutions or errors going unnoticed. People going against the machine's proposed outputs puts them in a difficult spot because a person cannot rationalise the process that the AI used to create the proposed solution. So, if a doctor goes against the advice of AI on an X-ray examination, who would be liable for error? Although the broader patient health context is still critical to reaching a conclusion on any appropriate course of treatment, it is therefore reliant on the social and community aspects of the doctors problem-solving.

The human-human social bond between the customer agent and the customer is critical to opening a healthy dialogue. This enables the successful analysis of a customer's issue. These are the tools to develop the customer service person's skills to have an effective and fulfilling customer interaction vs that of the soul-destroying, inhumanity of speaking to a machine.

People may be deployed as second-line troubleshooters to resolve the issues that AI can't after a long series of filtering and menus. At this point in the process, the customer is angered because the social bond is lost, so the customer agent doesn't get the same degree of information from the dialogue with the customer. This limits their ability to adapt and develop their skills in a variety of circumstances. Customer service representatives need autonomy and accountability to make decisions because that accountability transforms the motivation of the representative, which in turn drives the value of and the overall customer experience.

The solution is to train your people and give them autonomy to achieve a win-win scenario with the customer. AI does make customer service people less effective by demotivating teams because management are endlessly looking for additional ways for technology to take people's tasks. This destroys human value, capability and skills development. To what end if the outcome is only achieving like-like productivity alongside an inhuman, soulless customer experience?

Balanced experiences and lessons about human intelligence, value, motivation, and adaptability are completely ignored by the AI community. Thinking such as “people are machines”, “more tech will fix it”, etc, creates blind spots. These flawed mantras justify the technological and AI drive strategy to the detriment of business performance and customer satisfaction. Technologists will, therefore, be blind to the role and contribution of human skill and how to develop and harness that.

Source: Jordan Peterson

Interestingly, Geoffrey Hinton Turing, award winner and Google scientist suggested that “people should stop training radiologists now”. That creates a blind spot. He goes on to say, “within 5 years deep learning is going to do better than radiologists.” The reality or strategic disconnect is that radiologists’ diagnosis requires even more situational and social intelligence than customer service, as does engineering design, autonomous cars or anything else that has a critical impact. These solutions require a tremendous situational intelligence to adapt to changing or different circumstances and to take cues from inputs such as pedestrians, patients, and other drivers.

Alpha Zero’s chess playing was able to come up with solutions in different scenarios or create solutions to varied problems, but neither Alpha Zero nor ChatGPT showed any judgement. You need judgement to ascertain whether outlying data is an error or true or this leads to inaccurate predictions. Data scientists use overfitting to ascertain animals in pictures by considering the context, such as the background. Yet when habitats change, wrong predictions will be made. This creates a false sense of success when, in fact, the AI is performing badly, and the output is at risk of being false and unchecked.

AI has no social intelligence, so it is completely dumbfounded when it is trying to make sense of an inherently social situation because the cues are constantly changing. Automation reduces rather than increases flexibility. Well-trained workers will fluidly respond and use their full gambit of peer-led community skills and resources to judge right or wrong and solve the problem with a robust connection and social bond with the individual concerned.

This positive social interaction maintains a relationship with a customer. This relationship is a critical element to creating an experience that is hard to replicate and or create a genuine differentiator or source of profit. When everyone else is blindly implementing technology for efficiency or productivity gains, they don't realise that they are trashing the value of the social bond. Such a relationship and experience reduce the perception of a purely transactional experience, which makes the product or service purely price-centric and devoid of loyalty.

The practical motivational implications of workers or professionals interacting with customers are driven by their degree of autonomy, authority to make decisions and their decision-making boundaries. These factors motivate staff and thereby drive customer experience. What is it like to speak to a person dealing with your important issue who has no authority or responds to commands and prompts in a standard, inhuman way?

Automating tasks leads to so-so productivity gains, devoid of a social bond, and a demotivating customer experience. This is a downward spiral of employee motivation because it negates the human aspect and any respect for human excellence in carrying out their tasks.

The marketing power and narrative of “Trusted AI”, “FOMO - don't miss out”, and “don't get left behind” drives these corrupting strategies and reduces company profits with the baseless promises of future productivity gains through the inappropriate applications of such technology.

Technology has good, narrow applications such as vehicle routing, where the complexity of routing over ten drops for a driver runs into billions of permutations or satellite navigation systems, which can make sensible decisions based upon real-time single-dimensional data alongside a person without reducing their effectiveness. But if a driver doesn't need to think through and learn a route, the driver will lose their ability to devise or check a route over time. In the same way a mechanic who immediately refers to YouTube videos will stop using and honing their skills to develop their intuitive hypotheses-based analysis of the issue to inform the course of action. So even with these narrower applications of AI, there is a cost to human value and maintaining their capability.

Machines adopt a completely different process for solving problems, such as ChatGPT operating word by word to propose the next word to form sentences; we have no chance of being able to keep up or check the validity of the output. The issue therefore is whether humans will see, stand up against and speak up against the AI systems' false outputs. And will people's skills in this environment diminish because people will also be manipulated into thinking the same way as the machine, and thereby undermine their own judgement of what is right or wrong or follow their own intuition? This will be very hard to articulate at volume or constantly when working at speed. For an operator to constantly say no would require a huge amount of courage to stand up against the perceived technological wisdom (or maybe stupidity) and likely lead to operators' capability being challenged.

Companies like Amazon don't value people as human beings. Regardless of an Amazon employee's pay, there is a fundamental sense that Amazon jobs are not good jobs because workers must abide by strict, fast-paced routines whilst being constantly monitored without any autonomy. This destroys the employee's potential value, employer-employee trust, uniqueness, or humanity. Terminations can be automatic based upon the data collected. This is treating people like robots who are being monitored by Amazon's automated systems.

Amazons or others' technological focus creates non-thinking and race-against-time environments. This causes unsafe working environments and twice as many accidents as other warehouse businesses. It's dangerous because employees are not thinking and are dehumanised, so they almost certainly don't care what they are doing and will have no goodwill towards their employer. This is a seedbed of an environment where those who care will look actively for opportunities for sabotage. They are probably more justified in doing so than their forefathers, the Luddites.

Amazon's monitoring of employees trashes the value of their employees' humanity and enables them to cut wages. Monitoring is an activity used to prevent the sharing of productivity gains, increase Amazon's rents upon workers, and break the link between the potential value these employees, given the chance, could bring.

The trend of elite to automation and surveillance leads to a type of work that is now created to respond to shifts in demand. So, we now have zero-hour contracts, no opportunities for promotion or headroom for growth, alongside extensions and cancellations without any notice. These are both symptoms of the so-so automation of roles where there is no productivity improvement but a multidimensional cost to workers.

The underlying narrative that is driving this must change to machine usefulness for people. The greater the disparity between worker's interests and the technology, the less the technological-centric employers will value their employees. So, fewer opportunities will become available to demonstrate people's value in surveillance-based workplaces.

Deploying an automatic machine or treating employees like automatic machines, regardless of their feelings or fragility, is the equivalent psychology of the slave trade.

There is a naivety and a systemic background process that will accelerate this change because people cannot keep up with the data processing and "decision-making" of machines. So we won't know when to switch it off. Yet people can reprogram themselves on a job based on changes and new information.

Kai-Fu Lee summarises the assumptions and dominant dialogue underlying these damaging strategies: “Robots and AI will take over manufacturing, delivery, design and marketing of most goods.” Industry 4.0 has enabled us to manage complex supply chains, yet if we focused on simplicity and people collaboration, we would achieve a better, more balanced outcome.

Japan values flexibility and practices worker participation in decision-making. What people now call learning is now forced upon them, yet everyone is different. In theory, I-based learning can be adjusted for small groups or bespoke for individuals. It was noticeable during the pandemic that the value of learning without a human context, relationships, or social cues is debatable in terms of whether such learning “went in” or had any cognitive value in the absence of social conditioning.

The web has provided unparalleled access to information and can bring people together with different skills to potentially fuel economic dynamism expanded by digital technologies. AI is good at providing the information to better utilise assets, for example, by sharing information about fish catches with vendors or Airbnb expanding consumer choices beyond hotels. These in their own way are parasitic tech models that intermediate and control sales channels. They take excessive rents, not margins, from the service provider and leave few alternative sales channels for hoteliers. The crux of these business models for fishermen or hoteliers is the imposition of economic rent and digital feudalism without value add from the intermediator that controls the sale.

Human complementary machines that are intent on cost-cutting are not attractive. Yet 80% of the $30bn spend by tech giants on AI goes on surveillance and automation. So why are they not developing tools to develop capability? It’s because anything else except like-for-like replacement of roles through automation will create higher value roles, which will increase costs and the collective bargaining potential of the masses. This leaves the tech elite’s focus on cutting labour costs in already financially strapped schools rather than creating new higher-paid roles!

AI reinforces biases and runs on biases in its data or what it is instructed or programmed to do. Technology solutions must be bespoken to national interests rather than, for example, pushing generic GM crops on societies that face different pests, which are too capital intensive, require too high skills, and reduce their agricultural productivity. These crop technologies create a nefarious economic dependence on these inappropriate crops. Estimates are that inappropriate crops like this cost 40% of a farm's productivity, so increases inequality. The GM crop providers don't seem to care about local interests either. This is common with the psychology of digital technology in that these applications are implemented regardless of the local, human, or economic implications. So, GM crop-type technology often favours the technology provider and not the customer.

The success of Singapore's turnaround and rise from a third-world nation over the last 30 years was based upon pragmatism, what is good and honesty. New technologies in S Korea, Taiwan and China enabled human resources to be used more effectively, creating new employment opportunities, increasing investment in education and making a link between technologies and people skills.

So, rather than improving the lives of people or raising productivity and wealth, these technologies are creating demand for their own services, highly skilled workers, and capital intensity. This opens the door to inappropriate and skilled services from management consultants working in conjunction with the corporates, WEF types, investors and the tech elite. This selfish tech strategy shifts wealth by imposing rents on the masses to increase the wealth of the tech companies.

AI fuels this self-interest in the hands of the tech elite by creating more ways to automate work, sidelining human value, and doing deeds such as addressing the "*major problems facing humanity*". The tech elite are destroying the basis of humanity, the value of individuals, their unique value and the different views they bring!

This misalignment fuels the tech belief that "people are stupid" and systemically reinforces leaving people out of the consultation process because they don't know what is good for them! These paradigms are fuelled by the tech elite's ignorance of the value of people and what good looks like. The tech obsessions are driven by their underlying assumptions that "people are (incompetent) machines" and/or "more tech will fix it", etc

This marriage between technology and big business, WEF, Big Pharma, Big Funders: Blackstone, Vanguard, etc and others has created a growing number of billionaires and fortunes have multiplied once AI started spreading in the 2010s. These elite and bodies are interested in protecting their power and wealth by fuelling their related, often left-wing ideological beliefs beyond the technology itself. So ESG and gender ideology are also used to control the agenda between the tight limits of what they perceive as right or wrong. So the technology providers that own the communication and cloud of data and platforms naturally collude to control their agenda, reinforcing their common narrative at the cost of the unpaid cloud serfs that keep their platforms alive and up to date.

The fact is that AI and "digital transformation" (a misleading mantra in itself) have turned out to reduce productivity or create at best like-for-like productivity. This has destroyed the basis of shared prosperity, democracy, free speech, and people's ability to revolt against such strategies to influence strategic realignment. These organisations have as much influence and more financial clout than governments too. BlackRock controls assets worth 50% of the US's GDP. This is leading to fascist-type control and influence of governments. The UK Government uses Facebook and Google on panels going back for years, and the Chinese are influencing strategy in the UK Government too!

Technology therefore is no way to build shared prosperity or empower workers. The facts show exactly the opposite is true. It does enrich the tycoons, opens up the financial panacea of monetising information about people, and enables a new level of surveillance and manipulation.

The tragedy is that without adequate democratic checks, the more the technology fails to deliver benefits, the more rushed the tech elite drives its implementation. This closes the gap for people to catch up, dissent or ask questions. Many have come to believe and are duped by the mantra that "technology is inevitable". This is a lie because we always have choices. So technology spend is focused on automating work, escalating surveillance, and deploying AI to monitor and manipulate us.

So, we have two tiers: the big tech tycoons that know what is best for us and who will fix the big problems, and the stupid humans prone to errors and don't know what is best. The masses create the problems for themselves, so we are ripe for replacement.

This is sinisterly endorsed by the WEF mantra that the root cause of human problems is an "excessive population" (of these idiots). I guess the protagonist elite exclude themselves as the ones who should volunteer to jump overboard. Gender ideology too is aligned with population reduction by trashing the nuclear family and heterosexual relationships. Note there is no H in the 153 gender varieties LGBTQ. Q is the key disorientating questioning which keeps people in a confused state of lunacy-led uncertainty. Yet the facts are clear: you are simply either male or female.

Digital did not have to be used for automating work and undermining or destroying power or democratic influence in society. Unless the underlying psychology is addressed, "that people are too stupid to contribute" and/or the baseless assertions AI will resolve (rather than continue to exasperate) the world's big problems. And if the general mantras, such as "tech will fix the issues", remain unchecked, we cannot restore democracy. There is far more chance of casting in stone a path to a tyrannical dystopian social scoring system to control and monitor everything we can do and or can't be trusted to do.

Democracy dies in darkness and silencing fuels evil.

China and Russia are the propaganda experts to break down society. They know how to create a seedbed of constant manipulation and constant surveillance. China spends $6.6bn on monitoring and censoring content to the level of detail of individual and household habits, jobs, spending, and hobbies for inputs to predictive policing policies of the Uighur and broader population.

China uses a social scoring system to monitor undesirable and untrustworthy activity and subversive anti-government activity. These technologies are similar to the drift and data obsession that Western Governments are getting drawn into in order to keep up. China's system is implemented through technology providers such as Didi, Alibaba, Tencent and others. Chinese defaulters are punished by preventing free movement and fines, which form the digital dictatorship.

Government collusion is also common in controlling the West narrative: one response to COVID-19 was the UK Government used our military-grade propaganda tools to silence dissent. Russell Brand's deplatforming and demonisation was led by a UK Government minister. The WEF led the adoption of the UK Online Safety Bill. Our CEO of Ofcom is on the WEF panel that enshrines their anti-free speech doctrines into the hands of the tech companies. This outcome is exactly the opposite of the liberty people expect from the internet. The manipulation and control of narratives don't allow different perspectives to reach the truth through healthy debate, so reality is held in suspension.

So how did we end up with autocrats suppressing information and dissent and social media becoming the hotbed of misinformation? How are we being manipulated by increasingly authoritarian UK and Western governments in the same way as China and Russia? Once the mass data was collected on the population, these surveillance tools became incredibly useful to governments and companies, so they became increasingly interested in surveillance and manipulation. As people became more disempowered, top-down control has intensified in both democratic and authoritarian governments, and new business models have emerged. These business models focus on maximising engagement, vast data collection, and controlling attention through addiction. Outrage creates retention, so sick mental health and dulled people have become the norm.

Mao started off well, allowing criticism of the communist party. This was short-lived and evolved into an anti-right campaign, ending in the execution of 500,000 people between 1957 and 1959. Then there was the Tiananmen Square massacre in 1989. By 2020, dissent became much harder in China, with the internet in the hands of the authorities to sanitise and crush dissent. Xi Jinping also started with a softer approach in 2014. This ended in a hard line in Xinjiang. In 2017, a new generation of Chinese AI surveillance and cameras were developed. China's spend on AI now totals 20% of the global AI spend.

Furthermore, the Chinese population were fed on a diet of how China was superior, and propaganda criticised Western democracies. This led students to report higher levels of trust in the Chinese Government. The propaganda worked because students now deem the Chinese system to be more democratic by those subjected to the propaganda. Their views were highly influenced by this propaganda and led to unquestioning support and an unwillingness to listen to critical views of the CCP.

What Orwell feared were those who banned books, but Huxley was more concerned that there would be no reason to ban books because no one would be concerned or interested or cared enough to challenge the perceived wisdom. We are seeing the same population dumbing in the West too. There was virtually no revolt in response to the roll out of Sadiq Kahn's baseless ULEZ expansion. There was harsh treatment of those vandalising the ULEZ cameras.

In Huxley's scientifically manipulated dystopia, people grow up to love their servitude. So, they had no reason to revolt or push for overthrowing the dictatorship. Protests, such as the dissent in Hong Kong in response to Chinese rule, act as a trigger for an increase in government control and online control. Pegasus monitoring software can be installed on mobile phones as a no-click system to track activity. It is deployed in UAE, KSA, and Hungary. Jamal Khashoggi, the Saudi journalist, was under surveillance by Pegasus software before his brutal murder.

This focus on governments to deploy digital strategies leads to a focus on surveillance to suppress opposition and monitor citizens. This causes governments to be more authoritarian and intrusive over time. The introduction of the Online Safety Bill in the UK has already seen a direct increase in authoritarian control. Caroline Dinenage, Minister of State for Digital and Culture in the Department for Digital, Culture, Media and Sport, demanded the shutdown of Rumble in the UK for not following the UK Government Culture Minister's dictate to silence Russell Brand.

Early hopes of digital democratisation have been dashed because the tech elite has put its effort into influencing government in collusion with the WEF, WTO, WHO, Google, Facebook, and other fascist bodies. In other words, those bodies are dictating government policy, digital strategy and censorship.

Surveillance requires massive amounts of data, which requires technology and AI to process and make sense of. This massive amount of data cuts people out of the process, so as Yuval Noah Harari suggests, "Technology favours tyranny".

Once AI strengthens authoritarian impulses, it creates a vicious circle that requires an increase in data collection, which follows a reduction in human intervention. This makes AI a fully-fledged surveillance anti-democratic tool by cutting out people, and its roll-out requires increasing amounts of data, so it is an escalating catch-22 loop of more technology, less privacy, less human intervention, etc.

Huawei has exported its surveillance tools to 50 countries, giving them unrestricted access to these governments' data. This raises significant concerns about the nefarious influence of the CCP on these countries. That the USA has done the same is understandable to keep ahead because it's a race-bottom. The US's lawlessness in influencing social media is justified by the psychology of its "War on Terror" or any other such convenient mantra.

When talking stops, wars start.

AI-powered social media is a manipulative tool to undermine democracy, human rights, privacy and a means to provide top-down censorship. So, AI is likely to fuel revolt and "bad behaviour" in those who care, fuelling their cancellation. This leaves the weaker majority compliant, silent, and oppressed.

Violence is fuelled on social media: Facebook fuelled the genocide in Myanmar because a system like Facebook that is focused on maximising user engagement is fuelled by content that creates strong emotions. So, the intense engagement created through hate speech and misinformation related to the Rohingya was favoured by the algorithms. This triggered intense engagement from hundreds of thousands of users, so murder was orchestrated by the Facebook platform.

Social media platforms are motivated by controlling user attention, so they fuel radicalism fuelled by the compulsive language of hate speech, extremism, and misinformation. This enables Facebook and other social media to sell more ads. So, nefarious communication diffuses significantly faster than the truth. So Facebook's algorithms were elevating these sensational articles to create significant conflict, which led to genocide and murder.

Echo chambers are almost certainly an inevitable byproduct of social media because people naturally group around those who support their views. Echo chambers are not an environment that further reaches the truth. Echo chambers fuel the generalised views of those groups, leaving the facts unchecked. Echo chambers isolate dissenting views and reduce the skills or aptitude to engage with others who hold different views.

In 2015, YouTube engaged Google Brain to improve the platform's algorithm. Minmi Chen boasted that we can lead users to a different state. This makes the YouTube platform ideal for radicalising fringe groups. Celeb Cain describes how he became increasingly radicalised on YouTube, falling down an Alt-Right rabbit hole, falling deeper and deeper into this as he watched more and more content recommended by YouTube algorithms.

Controlling attention depends upon creating engaging content. But people respond far more voraciously to negative or outrage or indignant content rather than positive content. This content leads people astray and controls their attention by fuelling addiction or triggering dopamine. The data collected during this process is the basis for creating targeted ads. These ads are only as good as the attention that people pay to them. The vicious cycle of user engagement is self-fulfilling.

Why not let human choices guide how search algorithms prioritise content? The answer is positive content is less engaging, so it would reduce the engagement and addiction or ability of social media to control attention. So, positive content would reduce social media revenues.

Google's AI can track a user's IP address to create individualised ads for users. A detailed set of criteria to test and measure the effectiveness of this advertising hones it further and traps the user to influence their decision-making.

So why is there so much focus on AI? It's because it drives targeted ads and validates and refines the massive data collected on individuals. This entraps people every time they use the social media tools. The growth of these platforms is driven by user attention, so addiction has resulted from competing against other platforms for user attention. Social media constantly tests engagement against different types of content to refine the selection that is presented in users' feeds.

The like button was introduced in 2009 to act as an emotional cue for users' engagement and the appeal of their content. Likes fuel addiction, anxiety, inadequacy and envy, which all fuel low self-esteem. So those who were encouraged to stop using Facebook are significantly happier. The expression used by employees summarises the reckless mantra of Facebook employees centred on driving engagement "Fxck it just ship it". So, distractions or other considerations did not stand in their way.

Sacha Cohen said, "this is about giving some of the most reprehensible people on earth the biggest platform in history to reach a third of the planet". It follows that bad, false, or negative content that creates an intense emotional response will thrive. This encourages people to develop rage-creating content to increase engagement.

This engagement motive has pushed the tech industry in an undemocratic, manipulative direction. This polarisation and tendency to extremism limit the multitude of different voices from being widely heard. So falsehoods go unchallenged and good democratic policy decisions lack a healthy, varied democratic input or public debate.

The ability to freely engage without a strict tech elite-driven hierarchy or predetermined status seems to have been squandered, trashing the opportunity to create a new public sphere. This online public sphere, if set up without these nefarious types or negative stimuli to fuel emotional responses, probably wouldn't have been that engaging anyway. We are "designed" to cope with face-to-face, naturally limited local communication that triggers dopamine naturally through positive connection, healthy debate, and interaction.

Exposing people to such a vast series of connections with different views must be censored and/or contained with limited nefarious stimulus in people's echo chambers, or else escalating violence will erupt.

Democracy doesn't fit with the online rent-taking and engagement-focused tech business models and their AI illusion. In fact, it is in direct conflict with the tech elite's view "that ordinary people are too stupid to determine that level of decision-making", or the psychology is that "tech is the sphere of the genius's striving for the common good". From this (flawed) psychology, it follows that they, the tech titans, should be the ones who should and are making these important decisions.

Our historical social system, which has been the crux of society's wellbeing, cannot cope. When governments and tech elite think derogatorily about people, it follows that their data becomes things to be harvested and manipulated to achieve the "common good", so engagement is not something to be encouraged or in any way protected. In fact, democracy is a threat, so the free engagement of ordinary people should be destroyed, and dissenters marginalised or silenced.

The AI illusion favours the anti-democratic even though the elite describe themselves as centre left. This conveniently fuels the woke type of ideology that too imposes silence, tyranny, conflict and fear and the same type of self-righteous indignation for ordinary people who don't fall into line.

VCS believe their own elitist truths, such as: "people don't understand these technologies", so are "unnecessarily worried about the implications", and "Most of the fears are overblown or altogether unfounded", is common psychology. This shuts down the challenge or debate. So VCs forge ahead, accelerate plans and ignore these concerns because, like the tech elite, they believe that people will realise the benefits only when AI is integrated into ordinary life!

Disruption as a virtue and the "merits of connectedness" create blind spots justifying "the move fast and break things". This psychology has accelerated the undemocratic input and blinds the elite to others' views. "Disruption" has come to mean any negative impact on others or democracy, so disruption is fair game and to be encouraged.

Greater exposure to and normalising negative emotions and lower exposure to positive emotions impact users with lasting adverse effects and engrained habits. Google hasn't traditionally sought permission for proper consent, ignoring privacy concerns, so it acted first in the hope that the outcome would be accepted. Largely, this has worked out for Google. This leaves tech companies routinely collecting vast amounts of data without user consent.

As Hitler and Father Coughlin showed in the 1930s, inflammatory propaganda and rhetoric reaches millions. Now, such rhetoric can reach billions on social media. This is how authoritarian governments foist their vision on everyone else.

Wikipedia is a good example of an organisation driven by the crowd's wisdom. They don't follow the narrow interest of the tech elite. Wikipedia is not monetised using ads, so it doesn't try and monopolise user attention. Content is stewarded by their voluntary Arbitration Committee to monitor misinformation.

Seeking intense user engagement is conducive to the worst type of social interaction, polarisation, and mental health. To compete, the only option for ad-based revenue tech models is to push the psychological dopamine and serotonin buttons triggered by negative emotional reactions. They use their surveillance to fuel inequality and marginalise and atomise large segments of the labour force.

The tech elite and investors think the same. Although Elon Musk takes an interest in free speech on Twitter.

If everyone lies, the consequence is that nobody believes anything any longer. In a social media context, this leaves individuals atomised to believe their own truths and believe their own uninformed, unchallenged thoughts. As Jordan Peterson says, “our uncut thoughts are nearly always wrong”, and “if these thoughts remain unchallenged, they become our facts”.

With the cabal of colluding tech and corporate elite controlling politics, it’s almost impossible to build any organised opposition because the media channels and means of communication are corrupted and censored to prevent dissent. So, robust dissent will not rise in China because AI is choking democracy and providing increasingly sophisticated and manipulative tools to control oppression. This web of technological and media tyranny is prolific.

The Chinese education system and the effectiveness of military-grade propaganda tools in the West show that the mind is controllable. This leads people to stop trusting the news and become atomised, vulnerable to remaining trapped in their “manipulated information bubble”. Addiction to social media causes people to become atomised in their community and reject democracy. This alternative reality is created by persuasive emotional dialogue online that fuels extremist positions. This is unchecked by any balanced counterview or healthy dialogue, so it is at risk of being devoid of or cynical of the truth. In echo chambers, compromise is condemned. The more this takes hold, the more polarisation and atomisation escalates, and the more people become isolated in silos. This isolation causes social skills and mental health to suffer. Manipulating thinking causes stress, causes people to lose themselves and makes individuals increasingly socially isolated. This is psychological torment and undermines a person's ability to make a productive or balanced contribution to society whilst protecting the dominant elite mantra from going unchecked.

The metaverse takes this manipulation a step further. The metaverse is a system designed to accept this fallout and contain this breakdown of wellbeing, communication, and society that has been historically based upon productive people – people contact. The metaverse does this by subcontracting your communication to an avatar, so the metaverse just obfuscates and exasperates the issue, cementing people's marginalisation and isolation "in their personal pods". Along this tech trajectory, people's communications skills and increasingly entrenched ideological positions create escalating conflict, but that's in the metaverse, not the real world so contained. This process is a catch-22: more tech, reduced usefulness of people, more tech,

> Messing with a user's mood or memory is one thing. If online advertisements have the power to reshape a person's image of themselves, as that last Ohio State study suggests, it may be time to stop thinking about the internet as something people use, and start thinking of it as a new environment in which human beings exist and adapt, for better or worse.

reduced relevance of people, etc.

Source Unknown

Rather than address the cause by stopping the damaging aspects of this tech, the surveillance obsession, the Trojan horse of governments, and the obsession with collecting data on people will reduce personal privacy, as enshrined in the UK online safety bill.

The increasing power and influence of these tech, media, and fascist bodies, such as the WEF, WHO, UN, Blackrock-type investors etc, colluding to their own ends to control and protect their wealth and pushing aligned ideology makes protest difficult.

History shows that the only means of countering this level of tyranny is to revolt. But how?

Governments are trapped by getting increasingly drawn into joining the same race to the bottom AI and increasing surveillance, which requires vast data, infringing the public's privacy and rights.

The varied stream of arising crises holds people in a non-thinking state. Sophisticated propaganda and manipulation from the corrupted media and government lead to menticide or a non-thinking state. This silences the population and quells the will of strong individuals with the moral compass and courage to sufficiently care to dissent.

In parallel, the tech elite embed the leftist ideologies that trash the bases or ethical foundations of society, confusing the basis of identity, the biological reality of male or female, the value of a nuclear family, and the value of faith. Ideology such as gender, queer theory or critical race theory trashes the notion of truth, claiming that truth and science is a man-made construct.

This leaves the masses and youth floundering to find their identity or purpose and encouraged to make up their own set of weird and baseless ethics, fuelling the dystopia, confusion, and division. There is no successful, stable society that has ever been faithless.

There are underlying agendas driven by flawed climate and other non-heterosexual ideologies, whose plan to reduce the population conveniently suits trashing religious faith and the nuclear family, and the state, the elite are taking more control of the youth and their indoctrination, not education, so discouraging students to think for themselves.

Personal identities form the pillars of an individual's strength and society's cohesion. Organising a revolt now is tricky because people's identity is being lost. Personal or national identity gives them something to fight for collectively or individually. The tech elite control the media, which prevents collective bargaining. The leaders or informers of the "revolt", such as Russell Brand or RFK Junior, Trump, are being cancelled from traditional and tech media channels and criminalised. Technology has no time for personal identity.

People naturally look up to those with wealth, political influence, and status as role models. The mega-rich elite cabal are being less trusted, but naturally, "their success" gives them the right to determine what is right or wrong for the population. The tech elite and WEF-type organisations consider that only they can fix the big issues, regardless of their degree of democratic support.

The reality is that technology is having exactly the opposite effect on ordinary people, and the current tech trajectory only enriches and protects the wealth of the colluding elite cabals to the detriment of ordinary people.

Similar dominance has existed historically. Rockefeller controlled 90% of US oil refining in the 1920s. Now the investment for the first time in 50 years, has switched from oil and been overtaken by technology investment. So, investors are seriously invested in the tech elite's psychology too, so they support and fund accelerating results to achieve their returns almost regardless of the consequences. This is creating serious blind spots because the dominant tech psychology and bandwagon are quelling the tech elite's or tech investors' ability to reach the right decisions in their own echo chambers.

Wealth, innovation, and governance have only ever been productively enabled bottom-up by the masses taking many micro decisions. A few innovators create great ideas that turn into great businesses. The rest of the more stupid ideas fall by the wayside or are voted out by democracy. Right now, democracy isn't functioning, so these tech elite strategies carry on regardless and in fact are being accelerated unchecked to allow the stupid population to come to realise the benefits.

Woke ideology is weirdly defined as progressive or post-modern and has no objective check and balance or defined outcome. Woke is a set of attractive victim-centric beliefs that emerged from Marxism and the inequality of sharing economic power. When economic power lost relevance, Marxism switched to far more virulent, emotionally charged neo-Marx topics and groups to divide. These emotional, divisive groups are race, sexuality, and gender. Those who challenge the ideology get cancelled, whilst they define truth, science, or objectivity as a man-made construct. You can see that uncontrolled immigration and its violent social fallout undermine national identity and justify the need to quell violence using surveillance.

The outcome is a victim-centric mentality that seeks to control the dialogue and scapegoat those who fall out of their narrative and then terrorise "offenders". The outcome is silencing. Silence fosters evil. This evil is manifesting in gender mutilation and the trashing of truth and faith. This leaves conservatives without the skills or courage to defend themselves against these irrational ideological attacks.

Advocates of woke people work in our businesses, and EDI executives are appointed in corporations, universities and schools and this ideology has been infiltrating and subverting our education system for the last 30 years. Thirty years is the period that Russian propaganda experts know softens the population's will and leaves them malleable to any propaganda. Technology is the icing on this cake to police this tyrannical woke culture.

The mantras underly the so-called "progress" includes the WEF mantra of "climate emergency", "people own nothing and be happy", and the drift of government strategies to surveillance, climate change, and stupid populations. All lead to increasing control via social scoring.

UK Government collude with the tech elite to cancel dissent, and digital central currency will ultimately control our spending restrictions on movement, such as the UK's ULEZ expansion and 15-minute cities. The moves to switch to EVs will impact the automotive industry by trashing 800k UK jobs for negligible c0.01% global improvement in direct UK carbon emissions.

The blind spot is the indirect pollution and indirect carbon emissions coming from the coal-fired power in China making the EV batteries. The shift to EVs will hand ultimate control of the bedrock of our economies to China, which control the EV supply chains and the EV material supply.

Distributed wealth is critical to reducing pollution and providing the economic choices to take sensible daily actions. So, trashing our economies fits with the tech narrative and the smoke screen that is the climate emergency, when "climate emergency" is no more than a natural cyclical process going up and down that has taken place for millennia.

M | FINANCIAL TIMES

The former Ontario privacy commissioner says Sidewalk started with noble ideals and at first she got on well with Doctoroff. But with smart cities "people don't have an opportunity to consent", she says. "Technology is on 24/7. You can't remove yourself from it. Personalised data is a treasure trove . . . that's what everyone wants."

medium.com/financial-times/future-shock-inside-googles-smart-city-

Climate emergency, its negative economic status, and handing power to China is the ideal platform to accelerate the Great Reset that encompasses tech elite ideology, Chinese style surveillance, that is trashing democracy. These ideologies serve the tech elite or less than 1% of the population, dominating the agenda and controlling the narrative, rendering dissent almost impossible.

The tools used to quell dissent are far more pernicious now than in the time of Rockefeller when they controlled oil refining in the 1920s. Labour in the 1920s could organise dissent, albeit at a cost. In the 1870s, the strikers lost their lives when they protested about the Rockefeller-owned railroad and steel.

US presidents and UK Governments sided with these tech and industrial elite, creating a dependence which has created the current level of corruption by the military-industrial complex or Pharma. Led to the US Neo-Cons colluding to wage self-serving wars in Ukraine or control vaccine mandates, these are used to accelerate the tech elite and the WEF-type cabals plans.

What is the solution?

The Tilman Act of 1907 banned corporate contributions to federal candidates.

Large corporations remain powerful. Post-WW2, the New Deal reforms increased the sharing of prosperity.

More of the same isn't the solution. Addressing these issues is a psychological problem to unwind and identify the psychology that is creating these distorted, unfavourable outcomes based upon ideological mantras and untruths.

Ideology always leads to untruths because ideology is too general to be true, so it is false. Then ideology creates blind spots, so facts are searched out that fit with the flawed ideology whilst rejecting `counter facts or views. Ideology is a generalisation, so it is specifically too broad to be untrue.

The progressive movement is a fundamentally misleading term. Progressive equates to altering the narrative to change norms for good and bad and cultivating countervailing powers to change policy. It's clear that the woke progressive movement, as Trump says and can be seen playing out in the violence and corruption in the USA "Everything *woke turns* to *shit*… or just fades away". They control and redefine the narrative to suit their end; for example: "progressive paedophiles" wish to change the narrative from a criminal status to a "preference for minors". Because they are too young to consent, they should be protected.

The progressive narrative suits the tech elite because their strategies divide and immobilise, so they create infighting between the masses, rendering the chance of collective bargaining impossible. In the past this difference would have been the seedbed for robust, productive dialogue in a non-tech, democratic system, or when the dialogue ran out, violence or war.

Without an ad-driven revenue-generating structure, there isn't a viable alternative business model.

So, it's time to get back to basics because the tech industry is parasitic. It will kill and feed off any productive industries and their revenue streams. In the automotive, hotels, and banking sectors, the sale intermediators such as Facebook, Airbnb, etc, fuel unprofitable destructive growth-based models that the Chinese love and Western investors pump valuations then dump.

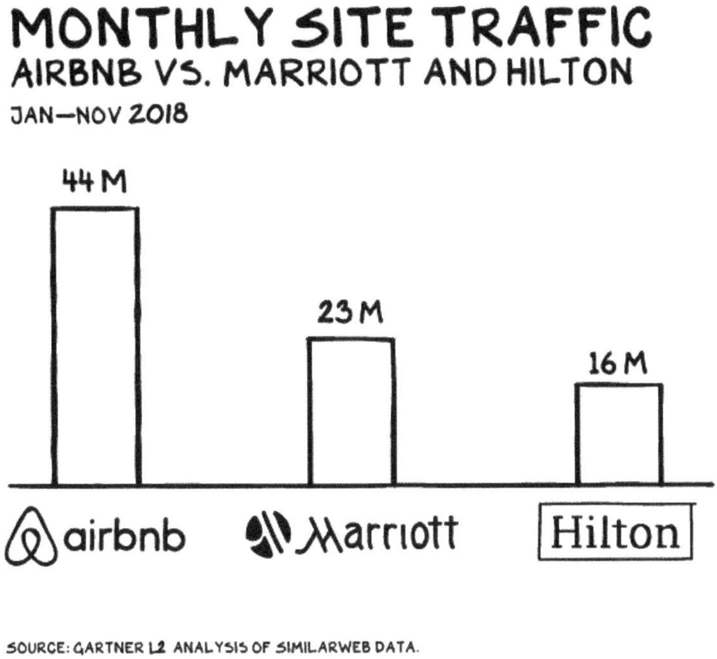

Greenpeace created the mantra of global warming in 1970 to counter the environmental damage of the oil industry. Al Gore fuelled a switch to cleaner products such as EVs, carbon taxes to reduce oil profits, and support for innovation and research into renewable energy.

These mantras lost sight of reality or any perspective, leading to the escalating "climate emergency" ideology. This has been fuelled by propaganda, colluding with the media and tech elite to control the narrative to escalate the crisis. This is all when human C02 emissions are only 4% of the global total. CO2 is a plant food, not a pollutant. So, the outcomes of a marginal increase in CO2 have been increased vegetation growing in the desert. Climate activists predicted exactly the opposite: an increase in desertification.

Vast subsidies for EVs in the USA have handed the US automotive industry into the hands of the Chinese, who control the EV supply chain. China has failed to make compelling ICE vehicles but dominates battery and cheap electric vehicle manufacturing and solar panel manufacture. The climate crisis suits China. Many who know the WEF read WEF as the equivalent of progressing China's interests.

The technological and associated narratives must change, but the tyrannical inhuman outcomes are fundamental to a technological system that at its centre is feudalistic and parasitic. Big Tech is inherently ad taking the cut or rent and profit from the productive industries that contribute something of value. The tech elite's incentive is to control attention, addict users to fuel its growth and protect its control.

Powerful gatekeepers, governments and corporations must stop aspiring to these tech titans' objectives. The data obsession fuels the demise of the productive industries and the wellbeing of people. Debate must be broadened because people do not have to be an expert in AI to make a valuable contribution to its direction. After all, who is tech serving the consumer, or for the first time, just the rent-taking tech elite?

The users of the tech elite's infrastructure are addicted. Addiction is an important criterion to understand. Addiction needs legislating against because addicted users lose control of their own actions. The addiction of the masses to technology and their manipulating cabal is far more prolific than heroin, with 15bn mobile phones in service and 2.5bn Facebook users being manipulated and driven to deeper, even more sinister dystopian systems such as the metaverse. So like "progressive paedophiles" wishing to change the narrative from a criminal status to a preference for minors who are too young to consent, the addicted need protection by the law.

Investors drive decision-making and their investment in this technology requires a huge shift in their focus. Yet, there are so few productive industries that can achieve such returns. The obsessive tech psychology and fear of missing out, and blind investors no longer doing due diligence properly. This leaves investors investing and believing in broken models like crypto, WeWork and Revolut. Revolut's cost of customer acquisition has exceeded sales for years. Crypto, fuelled by the mantras of being "digital gold", is, in fact, an asset that equates to literally absolutely nothing and its promises of "reducing inflation" and "handing back control of money to the masses" are just absurd.

Yet these business models carry on regardless, fuelled by baseless ideology and false future promises of success. The reality is exactly the opposite. These digital models produce nothing useful, reduce overall wealth and their rent-taking undermines entire economies and the people therein.

A small elite and a dwindling middle class are not a foundation for democracy or prosperity. This group creates a focus on reducing wages and labour costs under the mantra that the purpose and responsibility of businesses is only to "make profits for shareholders."

A unipolar approach to reject technology needs thinking through because China lead in the development and investment of AI. This creates a fear of the West missing out BUT also fuels this dystopian tech elitist race to the bottom. This race to the bottom only suits the Chinese Communist Party and preserves its self-interest and unaccountability. The CCP are above their own laws and Xi's purpose is to ensure the CCP survives at all costs.

A serious investment in the West need not follow this dystopian strategy. By restoring democracy and free speech, we would see a revolt, which would provide a chance to reverse this dystopian tech trap that suits no one. The assumption that many corporate executives that form part of the WEF/cabal organisation are probably making is that they want to be on the tech elite side of the fence. There is only room for a very few players on that fence, so the corporate execs, their own tech elite cabal-centric strategies, and their families and communities they live in will for sure cop the tech elite's fallout too.

Democracy, worker councils, countervailing powers and a different narrative should start in the context of the driving psychology to counter the mantra that "tech is inevitable" and "more tech will fix it".

Tech is destroying society, democracy, individual freedom and our productive economy, so let's wake up from this hangover and stop drinking the tech Kool-Aid because it is poison.

The fallout from the climate change ideology is overwhelming, but 100% suits China. The reality is that fossil fuel consumption, despite $5tn investment over the last 20 years, has only reduced 2% and 97% of cars are still fuelled by fossil fuels. To achieve a switch to electric, 35M tonnes of green metals must be mined per year, and 6.5bn tonnes including aluminium and steel. This requires 382 mining projects at the feasibility stage to be completed by 2030 to meet the demand in 2030. This would cost $300bn per annum just in capital expenditure to 2030. No one has budgeted for this.

There are only 500 cobalt, lithium and copper mines operating at present, and a mine takes 4-7 years to become productive. Environmentalists are increasingly, and rightly, blocking these mining projects. So, projects are being started in dicier locations, where environmental concerns can be brushed off more easily.

The only way this capex and infrastructure investment can be funded is through higher prices, so when cost is an issue now, as time progresses, no one will ultimately be able to afford to own an EV.

Ideology creates the preoccupation that these flawed strategies will somehow come true or that more technology will magically resolve these fundamental issues. Electric cars and hydrogen and tech-driven carbon capture are climate falsehoods because the basis of physics, or the end-to-end energy required, is uneconomic. The threat these strategies impose to energy security and the affordability of these new solutions renders them destructive no matter how much you develop these technologies.

Battery manufacturers are already using less material. A typical EV battery contains 69kg vs 80kg of copper in 2020. Abundant materials may win in the end, such as sodium-ion batteries and silicon anodes. Manufacturers may build short-range vehicles with portable batteries, which may provide an increased range of 600km, so it is rarely used. The industry needs vast increases in recycling and design for recycling and another $15Tn unbudgeted investment in global electricity infrastructure by 2050.

The ETC expects shortages of market-breaking magnitudes by 2030, a shortage of 10-15% of copper and nickel, and 30-45% shortages of other battery materials. This will kill the green transition and/or lead to escalating prices. China is laughing as they control material supply and, mining and processing. A major automotive manufacturing switch to China would kill the Western auto manufacturers who, at that stage, having made huge losses consistently on EVs, are already very fragile!

Alongside the pollution, the hidden infrastructure costs in the Indonesian nickel mining industry run at $20bn! None of these peripheral costs are factored into the cost of running or buying an EV.

Governments don't have this amount of money, even in the developed nations. So, this strategy will likely escalate costs out of travel, out of control, along with major supply shortages, but allow the WEF to achieve their objective by trashing our economies "of owning nothing and being happy".

There isn't enough lithium or copper or water on the planet to serve this market, so mining needs to increase by 1000%. Putting it simply, that level of mining will trash nature or the very environment we are looking to protect.

Environmental strategies are bespoken to countries. Broadly, the answer lies in moving away from the ideology and thinking of hybrid solutions.

EVs will play a small role in inner cities to reduce pollution, but ICE vehicles are still a very sensible long-term solution for many of the remaining journeys.

The EV industry doesn't create enough profit to sustain the economy as a replacement for ICE vehicles. With 50%+ of the value add making batteries in China, the industry is therefore controlled by and handed to China. In the meantime, China is avoiding the climate change objectives and continues to build thousands of coal-fired power stations to make these EV batteries.

Fossil fuels need preserving. Oil is liquid gold, and contrary to the false climate ideology, it must play an increasingly critical role in building the wealth of developing nations. The wealth enables the developing nations to switch their economies to their own set of responsible environmental activities. People without money don't care about the environment. The poor's priorities are simple; the day-to-day focus is to feed their families.

The West climate priority should be to invest their money in protecting the developing nations' economies from the natural cycle of sea level rises. They can learn from the Dutch and others in that and protect nature. Nature processing 96% of the global Co2 goes unmentioned.

Nature will remain the only means ever of protecting our natural environment and the critical species that all play a part in the ecosystem. We will never be able to understand the complexity of nature because nature is God's creation. As the bible says, the earth is our resource and up to us to look after it, BUT not fix it; that is beyond us, God will fix it.

We need to redefine with an open mind what good looks like. For example, Botswana has been the most politically successful country in the last 50 years. We need to find the courage to stop doing these stupid tech things and do the right thing. Manage the geopolitical and social risks from China rather than taking their offer of Chinese shortcuts.

The sheer volume of communication, different views, and disparate needs online create an environment that is fundamentally unmanageable. Humans have only ever maintained relationships locally with people like them. So, censorship is a fundamental requirement. This suggests that a tech-driven online solution is the wrong solution.

Application risk should be considered narrow because AI applications like video conferencing and satnavs have a downside but present little risk.

The psychological manipulation operation driven by AI dominating social media makes Gen Z overly fragile. When the next crisis arises, they may be so helpless that they will simply turn to the colluding government and associated tech/industrial cabal and helplessly give up all of their rights.

People need to stop using tech because they are using their weapon against themselves that confuses and destroys their identity, uniqueness, purpose, free will, and willingness or ability to dissent. Tech elite messages are dressed up with ideological woke, climate, health, and any other lies to make them taste good.

A platform devised by Taiwan called g0v allows individuals to create alternative versions of Taiwan's bureaucratic services.

Alongside reinstating democracy, the grip of big tech and its power over data collection must be broken up to increase the share of wealth distribution and shift its direction to increase competition and alternatives. We must concurrently protect ourselves from nefarious, more egregious Chinese tech alternatives.

Taxation is 20% too favourable towards technology, which disadvantages labour, so concurrent investments in labour should be encouraged.

The government must get the balance of regulation, taxation, and agenda-setting right to provide the open and democratic means to shift away from parasitic tech and towards more productive, fair, humane alternatives, and fast.

The government must also quell their obsession with data collection and surveillance, which undermines the privacy and sovereignty of the individual. Government must serve the people, not the other way around. The US Constitution is a good model as a starting point, although the UK created the foundation of the citizen-centred sovereign democratic model and its legislative system.

Our thoughts or psychology control all we do. We need to develop the skills to understand how narratives such as "tech is inevitable, climate disaster, hydrogen economy, green economy, green hydrogen, carbon capture, anti-oil movement" etc are controlling us and how we can become aware of the psychology to get control of our own and the dominant narrative.

The skills to get control of our thinking, identifying, and removing these strategic dislocations is the essence of the Winningthinking® game changer experiential learning program and supporting workbook "We have infinite capability".

Digital advertising and wealth tax should be introduced to shift tech businesses away from ad-driven parasitic services and towards other business models if a viable non-rent-seeking alternative exists!

A macroeconomic model predicting the impact of the crazy tech government strategies should be devised and checked in a democratic way with the public to evaluate the risks and options. We must rule out the stupid, unrealistic options such as a universal basic income or minimum wages. If the economy is trashed by these elitist policies, no one can afford this, nor will anyone pay for it!

Ideology in education must be removed to encourage pupils to think, take ownership and act.

Playing God in its various tech and elitist manifestations never did and never will work out well, so we must kybosh the hubris of "AI will solve everything".

We must encourage thinking: how can you tell if someone is thinking?

We must increase interpersonal human-centric connection and intimacy between individuals to create logos, the freedom to speak, value individuals for who they are, and the diversity of their views and reward merit.

An ethical system must be based upon sound spiritual principles that define what we perceive to be right or wrong. Humans are primarily spiritual machines, not machines.

We are not capable of defining our own purpose. The ethical solution is rooted in Christianity and Jesus Christ, our creator. Yet good and evil runs through each human heart. The alternative is trying to define the principles ourselves.

Look around you. Self-reliance and wokeness have caused the collapse of the USA. Self-reliance gives rise to diabolical consequences, sexual immorality, child mutilation and re-aligning the narrative to evil ends such as decriminalising paedophilia and thereby dumbing the narrative to minors-attracted persons. That is evil and escalating and normalising evil is the outcome of silencing people.

MIT Technology Review

It is the summer of 2023, and Rachel is broke. Sitting in a bar one evening, browsing job ads on her phone, she gets a text message. Researchers doing a study on liver function have gotten her name from the bar's loyalty program — she'd signed up to get a happy-hour discount on nachos. They're offering $50 a week for access to her phone's health datastream and her bar tab for the next 3 months.At first, Rachel is annoyed at the intrusion. But she needs the money. So she nods at her phone — a subtle but distinct gesture of assent that is as legally binding as a signature — and goes back to her nachos and her job search.

But as the summer wears on, Rachel can't help noticing that she's getting rejection after rejection from employers, while her friends, one by one, line up jobs. Unbeknownst to her — because she didn't read the fine print — some data from the research study, along with her liquor purchase history, has made it to one of the two employment agencies that have come to dominate the market. Every employer who screens her application with the agency now sees that she's been profiled as a "depressed unreliable." No wonder she can't get work. But even if she could discover that she's been profiled in this way, what recourse does she have?

medium.com/mit-technology-review/its-time-for-a-bill-of-data-rights-l

China's legal system now operates on law-based governance, enforced by professionals who work within written codes. Yet, the underlying ambiguous legal language makes arrest and fear of arrest more likely. The tyranny and state terror retains control and is escalated by using ambiguous narrative and arresting people as they see fit for thought-based crimes. Xi's new laws punish those that "harm the spirit" or "hurt the feelings" of the Chinese people.

Ambiguity In China's law has wider implications for what people wear, symbols, what anyone deems offensive, news, insults to party heroes, etc. These laws are decided by a single officer and have included a woman detained for taking selfies wearing a Kimono. A Kimono is a symbol of Japan, a country China loathes. When faced with harmless ridicule, a police officer may feel as though they have been abused based on their feelings. Chinese judges used to ask for bribes, but now they too are too scared to ask.

The will to do the right thing has to be backed up with action. One of the top 3 most effective actions to protect the environment is to protect the African nations from climate change. African nations make up 16 of the top 20 countries most vulnerable to climate change from the effects of extreme climate events. A pledge of $100bn per annum was made for this purpose in 2009 yet it has still to be implemented.

Chinese demographics suggest that over the next few decades, the working population will decline so fast that their economy will shrink by double digits. So maybe they are biting off more than they can chew. The debt pile is forcing the Chinese Government to tighten their budgets, so who will pay for this pointless climate-related plan exactly? Climate plans will exasperate wealth distribution. The lead indicator of whether we can act is our ability to pay, so we will find the global investment required to build this power, grid and mining infrastructure estimated at c$15tn per annum to 2050.

However daft the environmental strategy already is, the way it is implemented requires the global entities and governments to run very smoothly. Or we could face enforcing rigid tyrannical measures that could vastly reduce our current freedoms and trash our economies. All in the name of a pointless exercise to avert a non-existent crisis. Climate change is just a normal phase of a climatic cycle that has occurred since time began.

What can you do?

In the meantime, we must stop this amoral tech elite narrative and government-fuelled ideology driving these strategies, or else they will carry on regardless of the cost. We must develop the communication skills to maintain dialogue with disparate global players, governments, and corporations to harness democracy.

Let's individually play our parts and take ownership to refine this crazy strategy. Ultimately, change will be delivered by You and I.

Imperial powers don't leave a successful legacy, nor do they educate the masses. Imperial powers create reliance, not convergence.

We now could destroy the world or blow it up. Created by genius, used by all.

We must be social.

"So, let us not be blind to our differences—but let us also direct attention to our common interests and to the means by which those differences can be resolved. And if we cannot end now our differences, at least we can help make the world safe for diversity." JFK

Those words to end the Cold War probably cost JFK his life.

Princeton economist Anne Case and her coauthor Angus Deaton have identified what's likely a related trend. They found that mortality is rising among middle-aged white people in the US with a high school diploma or less. The culprits: high rates of suicide, drug addiction, and alcoholism, which Case and Deaton call "diseases of despair" because they don't seem related to poverty per se, but rather to disappointment;

Automation might be partly to blame for these social problems. But if economists like Acemoglu are right, the key to creating more good jobs is not fewer of these advances but better versions of them that are deployed faster throughout the economy.

Pittsburgh reborn

That, in essence, is what Pittsburgh's attempt at reinventing itself is about.

Source: MIT 2019

Young people are getting stabbed increasingly frequently for trivial things and reverting straight to violence rather than any discussion whatsoever. The stabbings in London between young people continue to rise with the breakdown of social skills, the absence of the male role in a nuclear family, lack of meaningful work opportunities, hopelessness, subversion, and atomisation of the young trapped on social media, and the inability to have a challenging discussion or the inability to cope with woke dialogue, and hopelessness-fuelled drug use.

The cause in a large part of this is the inhumanity that comes from increasing technology use and its overbearing control, creating real anger in men. Part of the WEF adverts for the Great Reset show the total control of society using technology. Paradoxically, the technological breakdown of society fuels and supports the WEF strategy; therefore, with this societal breakdown, more technology and surveillance are required.

Ethics and good behaviour come with accountability, freedom and a society that encourages the right spiritual values and creates worthwhile opportunities that lead to productive outcomes. This tech is in the context of the WEF, creating the largely baseless climate fear in the young as an existential threat. The outcome of this accelerating trend is to impose total control on the population, 15-minute cities, control spend, travel, social scoring, mediated by digital currency, etc.

> Princeton economist Anne Case and her coauthor Angus Deaton have identified what's likely a related trend. They found that mortality is rising among middle-aged white people in the US with a high school diploma or less. The culprits: high rates of suicide, drug addiction, and alcoholism, which Case and Deaton call "diseases of despair" because they don't seem related to poverty per se, but rather to disappointment;
>
> Automation might be partly to blame for these social problems. But if economists like Acemoglu are right, the key to creating more good jobs is not fewer of these advances but better versions of them that are deployed faster throughout the economy.
>
> **Pittsburgh reborn**
>
> That, in essence, is what Pittsburgh's attempt at reinventing itself is about.

Source: MIT 2019

The WEF ideology is tyrannical and a catch-22 that will fuel escalating violence.

Freedom, Biblical values, environments that create worthwhile opportunities, productive, not rent-centred businesses, and a shift away from this tech-driven dystopia are critical, or else societal breakdown will escalate.

Medium Social Media

To begin to imagine such a world, suppose that researchers could kick a sample of people off Facebook and observe the results. In fact, several teams of scholars have done just that. In January Hunt Allcott, of New York University, and Luca Braghieri, Sarah Eichmeyer and Matthew Gentzkow, of Stanford University, published results of the largest such experiment yet. Those booted off enjoyed an additional hour of free time on average. They tended not to redistribute their liberated minutes to other websites and social networks, but chose instead to watch more television and spend time with friends and family. They consumed much less news, and were thus less aware of events but also less polarised in their views about them than those still on the network. Leaving Facebook boosted self-reported happiness and reduced feelings of depression and anxiety.

source the economist

Parasitic, feudal, rent-taking digital companies do not create worthwhile opportunities. They starve society of resources, opportunity, and control. Tech business models must be regulated and taxed accordingly.

> One of my biggest recommendations is the formation of GAIA, what I call the Global Alliance on Intelligence Augmentation. At the moment people around the world have very different attitudes and approaches when it comes to data collection and sharing, what can and should be automated, and what a future with more generally intelligent systems might look like. So I think we should create some kind of central organization that can develop global norms and standards, some kind of guardrails to imbue not just American or Chinese ideals inside A.I. systems, but worldviews that are much more representative of everybody.
>
> Most of all, we have to be willing to think about this much longer-term, not just five years from now. We need to stop saying, "Well, we can't predict the future, so let's not worry about it right now." It's true, we can't predict the future. But we can certainly do a better job of planning for it.

medium.com/mit-technology-review/why-a-i-is-a-threat-to-democracy-and-what-we-can-do-to-stop-it

. . .

The solution to reverse this scenario requires humility, grace, forgiveness, and love. Jesus is the role model and master of these characteristics.

Thank you for reading thus far.

This introduction sets the scene, in the context of the following 5 books in the series for executive leaders:

- THE EVIL OF SILENCE: REMOVE WOKE CULTURE – the mechanics of division and tyranny – understand the psychology of envy, victim hood, terror of cancellation, loss of merit, forced thinking, and the evil in silence
- "EDI MUST DIE" WHY CANT YOU ESCAPE THE IDEOLOGY – EDI, WEF, ESG are the puppet master ideologies – understand why ideologies lead to escalating lies, loss of self and corruption.
- WHY CANT WE STOP? AVOID PROPAGANDA AND MENTICIDE – the 2020s cause of acting and thinking recklessly – understand how and why hysteria corrupts strategy
- THE PSYCHOLOGY AND PATHOLOGY OF BIG-TECH: THE TRAP OF PARASITIC TECHNOLOGY – reducing you and your business to dope on a rope – understand how tech fuels minutiae to quell debate, terrorise and create division
- THE SOLUTION: YOU HAVE INFINITE CAPABILITY – how to regain what good looks like

THE WAR ON OUR BRAINS FROM 2020-2024

Impairing people's ability to think through vaccines and propaganda puts society in a position where we cannot aspire to the obvious or real, rather must solve the "urgent" problems. The constant states of emergency and health and physical wars may not even be seen to achieve the Great Reset but a war against the human brain, for the technocratic elites to permanently secure the technocratic domination of the world.

This manifests itself as a constant stream of seemingly arbitrary restrictions and incomprehensible over reactionary strategy

If people are increasingly fuelled by fear and neurologically damaged by vaccines, they will increasingly become unable to think, so they will be increasingly infuriated by those who can still think ...

"What the herd hates most is the one who thinks differently; it's not so much the opinion itself, but the audacity of wanting to think for themselves, something that they don't know how to do" Arthur Schopenhauer (1780-1860)

To think laterally is a dirty word, to think at all is dangerous. The 9th commandment of the World Health Organisation states their uniform technocratic control over the traditional media narrative, which they did very well during the plandemic.

Man not only kills over food, but also power, ideologies and beliefs. Covid 19 vaccines have the highest fatal adverse events of any vaccine on the market according to VAERS, yet only 1% of all serious vaccine events are likely to have been reported to VAERs due to underreporting, so there may have been many millions of deaths from these injections in 2021 alone.

A 3-sigma event would give you a 10% increase in deaths or a 1 in 200-year event, so a 40% increase is just unheard of. Yet Penny Mordant dismissed Andrew Bridgens debate in the House of Commons a conspiracy theory, more below.

The secret boss of the WHO is Bill Gates having given $2.5bn since 2000, so he manipulates WHO policies with disastrous consequences. Once the WHO treaties are signed that would make Gates health dictator, so the WHO must be classified as a dangerous organisation without appropriate governance. Neither does the WHO invest in health improvements, rather it tyrannically reforms governments and dismantles the freedoms and rights of citizens.

Rather the technocrats are not managing health they are creating an operating system to monitor and control every aspect of an individual and every global community at will, with China their preferred partner. The WHO/WEF is an operating system based upon cruelty and inhumanity without national borders, cultures organised for the benefit of the technocrats – a tyrannical superpower that is our deadly adversary. The WEF target an 86% reduction in population to reach their "ideal population" of c500m.

A dystopian society controlled by AI where human culture will not exist anymore are also known as Smart Cities. Demos Helinski says a "Smart City is the internet of NO things"; no choice, efficient, post ownership, post market, post energy, post voting – because by knowing exactly what people want and do, reduces the need for elections, because behavioural data replaces democracy as the social feedback system.

Nonconformists are a real threat to such a system, but the erasure of cultural identity is not enough to eradicate personal identity, this would only succeed if personal memories were replaced by the narrative of the technocrats.

Such a thing has been made possible by the covid 19 measures, so targets non-conformist reprogramming and depersonalisation maybe even be at the core of the Great Reset program.

Without mental influence, the loss of ability of people to think for themselves and technocratic narrative corruption the Great Reset would be impossible. The intent to cancel people, could be softened by maintaining the ability to keep and manipulate nonconformists online, because offline trouble making nonconformists, are probably harder to monitor and control.

The fortunes of the 2700 global billionaires increased by 6% of the worlds global output, over the pandemic in the first year alone. At the same time global output collapsed by 3% so the benefit was grossly disproportionate to these technocrats.

The Revolt was published in November 2023, since then the manifestation of the UN 2030 plan and the link with the Covid vaccination program has become clear so is worth sharing.

Do we have destruction or creative destruction? What follows destruction of the "as is", maybe far worse. Take the Bolshevik Revolution of 1917 that was ultimately replaced by Stalin's totalitarian Soviet Union; so, it's possible that we are now witnessing ideological destruction creating more destruction.

Keeping the lies alive.

The lies are blatant, in the first 18m the wealth of the US billionaires increased by 70% reaching $5Tn. Yet according to Schwab all citizens of the earth should be stakeholders and benefit.

What fuels these ideological obsessions of climate and vaccines, it's the narrative, such as Scholz stating unless we finally break the ever-new mutations causing ever new mutants, we will cause new infections. This is impossible because a virus cannot ever be prevented from mutating, but that narrative justifies a never-ending war for all time. So, the creative destruction of Covid 19 may not end either.

The same is true of climate change "the case for going to war against climate change states "Once established as a war like threat, climate change can be addressed with the gravity, urgency and resources it deserves." And King Charles stated, "we must put ourselves on a war like footing, approaching our action from the perspective of a military style campaign" John Kerry said something similar.

States of war force conformity and total surveillance of human activity. "We must come to terms with the notion that there is no more privacy, our personal and professional data are gradually becoming fully monitored and therefore transparent" Schwab states. Central Bank digital currencies alongside this, could and would control all financial activities.

Science must be muzzled if it doesn't fit the technocratic narrative, and the propaganda machine is in the hands of the technocratic elite: Zuckerberg, Gates, Pant, Soros, Page and Brin.

In 2022 5.6m videos were removed from YouTube by google with 94% identified by computers not humans. Such as normalising the current nonsensical plan by creating false parallels:

> "Although the first seat belt laws met with strong objections when they were introduced 50 years ago, no one bothers to complain about such a commonsense measure anymore. In mandating vaccination against c19, governments can offer the same basic justification for protecting both individuals and society".

The justification for the Gulf war was weapons of mass destruction that didn't exist. The parallel of this mass destruction propaganda, is now from a killer virus both which deprived people of their liberties and disproportionate measures with deadly consequences.

Disproportionate yes because as Nehls advises, vitamin D is a great means of containing Corona virus by improving our immune response. But no-one acted, even though scientists virtually no-one should die from c19. It's not surprising that the vaccine protagonists and beneficiaries prevented people using vitamin D.

Social exclusion

The vaccines were not only misguided but also risks people's lives, and those that refused found it very difficult to earn a living under the false psychological justification of protecting society. Failing to fit chip implants could likely lead to the same type of exclusion from society too.

Chip implants are being developed by Gates and Chinas government could include payment and vaccination ID. Provided that these changes are drip fed so as not to spook the audience, and outdated truths can be deleted from the internet so that the technocratic narrative is adjusted when it no longer corresponds to reality. For example: the CDC altering the definition of what a vaccine is.

What enables the Great Reset?

The Great Reset is the nightmare of total control over human behaviour by limiting mental capacity using vaccines and standardised thinking achieved using the technocratic narrative.

The Great Rest is about expropriation, control, and total digital surveillance under the lies based guise or ideology of global health care and climate emergency.

It's not a response to an unwanted crisis but a long-planned takeover by a self-appointed tech elite. The takeover could not have been realised without a crisis and by infiltrating parliaments and subsets such as their WEF indoctrinated Young Global Leaders YGL".

The elites are so sure of their success that there is no need to operate in secret and they probably believe that serious resistance is out of the question based upon current conditions.

For a large part of humanity to fall into slavery, highly efficient brainwashing is required on a **global** scale and elimination of individual identity or unreflective mass conformity.

It's the role of the YGL to infiltrate governments to lead this transition.

If you were a tyrant with a globalist centrist intent to grab power and you could not achieve this nefarious outcome through persuasion, you would need to manipulate the masses to prevent a revolt?

The UN 2030 goal is obedience, unthinking conformity, through a very skilled, controlled selection of information, intensive propaganda, psychological coercive manipulation, and to threats of punishment.

Indoctrination is a vicious attack on our humanity, individuality, and freedom of thought.

The war on terror was justified by the 9/11 attacks. This made it far easier to legitimise military interventions, enabling many people to profit from the disaster and impose anti-terror legislation that is still restricting freedom and privacy today.

Dunning warned of the dangers of economic power and profit, at the cost and destruction of others, yet in 2019 the scale of and severity of Covid 19 measures and unbelievable profits made by so very few, and damage done to others using vaccines and imposed measures were beyond anything Dunning could ever have imagined. The super rich have benefitted extraordinary degree worldwide.

So, the super-rich Davos types, have ample reason and motivation to maintain the emergency to serve the larger goal “The Great Reset”.

Klaus Schwab boasts about the progress made towards the Great Rest during the plandemic.

Even Prince Charles is complicit, seeing Covid 19 as the great opportunity to reboot the economy, or implying “the install a new operating system” Yet how could this succeed, and the small elite get their way?

The same type of dialogue came from Trudeau, Olaf Schulz, and Olaf went on to say, “we should not be nostalgic for the good old days”. This is an intentional attempt to destroy our identity and history.

Yet fact checkers like to define The Great Reset as a “conspiracy theory”, thereby devaluing such discussions as childish nonsense. Yet a conspiracy theory only exists if people conspire secretly to increase their profits against the interests of others and by unfair means. So, the term conspiracy has been corrupted because that is exactly what is happening so not a theory.

The Covid plandemic followed a predefined plan leaving all governments imposing the same measures. Could one suspect intent given the super-rich increasingly controlled the media, politics, and science.

Following intensive fear-based media provided politicians an opportunity to present themselves at last as strong leaders and placate the fears of their neurologically compromised populus to avert (and abdicate their responsibility to the World Health Organisation / World Economic Forum) “harm to the population”!

Templates for dealing with the fallout were given to a small number of people to control how events unfolded. People were misled to think that the body’s natural immune system cannot cope with Covid infection which is untrue.

So, the experimental gene modifying drugs were imposed and justified using the psychology of “fear of death” and the “protection of others”. This ideology overran proper analysis and risk assessment whilst driving huge profits.

So, the psychology of “lifesaving injection” creates dependency and is hardly open to discussion, yet the vaccine made them more prone to pathogens.

So, the majority became helpless in the face of a tiny manipulative minority concentrating the power in the hands of a few technocrats, with complete control manipulating the interpretation of data and media, leading to a technocratically controlled state.

Zombie mode

People were forcibly unconsciously led into a reality of technocratic totalitarianism, which like amateurish woke ideologues, aggressively exclude dissenting views. They rely on destructive self-serving groupthink and the power to leverage the fear of exclusion. The common denominator between Islamism, woke ideology, parasitic technology and this technocracy explored in this book is the destruction of humanity and control, control of profit for the few at the expense of others.

Robert Malone describes this as "mass psychosis" and warns of the dangers of social landscape, media consumption, and dependence or addiction on manipulative technologies. This leads to mental enslavement and loss of freedom and us giving up our critical thinking.

Critical thinking relies on autobiographical memory which through technology, is diminishing more and more, so is losing its storage capacity.

This makes it almost impossible for people to operate outside of their normal routines even when their lives depend on it. So, people blindly follow trained beliefs aka "**zombie mode",** as opposed to thinking and planning new complex behaviours. This is caused by loss of autobiographical memory. With the loss accelerated by our alienated ways of life, covid 19 measures and spiking program disguised as vaccination.

Trapped in zombie mode it's impossible for people to critically question their own precarious situation, use natural curiosity, and interest in other course of actions are lost, opening the door to complete indoctrination.

Alongside the plethora of narcissistic gender and woke ideology, this leads to a reduction of emotional resilience leading to an increased fear of anything, but also a susceptibility of being controlled by fear.

So, identity is lost due to loss of functionality. Identity forming memories are replaced too by foreign content, propaganda or technocratic narratives, which leads to a mental conformity of society.

So technocratic narratives are about a “better world” because it is controlled by smarter people that “know what is best for you” and “even smarter artificial intelligence”.

“Welcome to 2030: I own nothing, have no privacy and life has never been better” Ida Auchen graduate of the WEF Young Global Leaders program

So, what are the points of attack that make people open to narcissism? Narcissism leads to a reduction in rational compassion, compassion is made possible by our social memory, which allows us to weigh up the consequences of our actions and question our prejudices. A nasty form of cancellation was experienced by anyone who refused to be spiked.

The longer this goes on, the less chance is there a way back from the Great Reset!

This can only be averted if one is aware of it, else the terrible result is the same: we are threatened by a zombie apocalypse, which is the end of self-determined creative culture, humanity, based upon individual freedom, identity, and creativity.

So, humans can be reprogrammed to be robbed of their individuality, so may accept servitude not just willingly but also happily.

Nothing is feared more by those that want to run the world than creativity and social awareness, so it is vital that these skills be revived and preserved.

People did not willingly give up their ability to think critically, so this dependence related state should, like heroin, should be legislated against because people have no control.

Indoctrination definition: noun “the process of repeating an idea, ideology, narrative, or belief to someone until they accept it and habitually enact it without criticism, discussion or question.

So, what was the issue with Covis 19?

Covid is a variant of a common cold for which it has been known since 2000 that there was no cure. In 2020 mRNA vaccinations had known serious side effects that would take 10 years to prove, but given the overwhelming fear created by the Covid pandemic, all measures taken were justified.

It seems clear that the Covid mRNA Vaccination was launched to impair people both physically and mentally. It would be perfect means of imposing a plan because the globalists "know better what is best for the people".

If you could stop people thinking, stop their brains regenerating, and use the centrist fear-based propaganda through a controlled media, to overwrite their memories, you could create an effective means to reprogram peoples' brains like autonomous beings with propaganda, leaving them unable to think and without curiosity.

Process 2 **CHINA!** used for producing the mRNA vaccines was not approved either and somehow created various amounts of fragmented DNA packaged in a way that could enter and corrupt our own DNA. This explains the sharp rise in cancer.

Physically, mRNA vaccines also once in the system continue endlessly to produce proteins, and the mRMA vaccines enter heart muscles and reproduce spike proteins in the heart and other organs leading to inflammation and myocarditis etc.

So, the issue wasn't with the threat of Covid 19 it was with the vaccines.

Yet the propaganda to reduce the population was usurped the simple action of taking their own action to improve their mental capacity by taking the simple steps to take Vitamin D and low levels of lithium, to enable the brain regenerate.

What was extraordinary, those advocating using vitamin supplements were accused of being Nazis thereby cancelled. Ivermectin which has turned out to be effective was trashed by being labelled as a horse drug etc.

There is evidence that roughly 50% of the population have a tendency that need looking after or are entitled in a rather socialist type of way. The rest are not that way naturally and take ownership for themselves.

The government actions taken shutting down the population and Covid loans etc, however created a socialist type of mentality by changing the habits and creating a reliance on the government for all. Such habits formed over 66 days become permanent habits, effectively rewiring the brains of the less compliant 50% of the population.

Governments seem to be trapped in an overwhelming disinformation / narrative bubble. This leaves the minority such as Andrew Bridgen who have spoken out against the vaccines and raised the issues of excess deaths and the validity of climate emergency, labelled as conspiracy theorists by Penny Mordant: Lord President of the Council and Leader of the House of Commons.

The accusation of being a conspiracy theorist was made when Japan took steps to ban mRNA vaccines, because of the explosion of cancer. Astra Zeneca and others are withdrawing their vaccines because of blood clots and other dangerous side effects.

The governments are not thinking either and probably are trapped in the most panic ridden bubbles of all, whilst trying to appease an increasingly polarised binary non thinking population, whilst indoctrinated themselves with their own fear of being cancelled.

Islamic ideology doesn’t naturally fit into this mix, but the common thread with the Great Reset is the Islam is anti-Western establishment. So, Islam is a useful tool to push immigration of Islamists into Western nations to destabilise the population.

The Islam threat caused a UK Government vote to be banned on Israel, for fear of Islamists reprisals for ministers if they voted in support of the real victims of the Hamas attack which was Israel.

So, ideology continues to corrupt the truth and enables even government Ministers such as Penny Mordant to avoid debate or challenging the prevailing wisdom despite the facts suggesting exactly the opposite.

The psychology of victimhood and worse third-party victimhood prevails with the bizarre twist of portraying the Hamas / Palestinians as victims when it was Israel that was attacked. And Israel has the right to defend itself against an organisation whose purpose is to eradicate the Jewish people. Yet in May 2024 the UN have just allowed the Palestinians to join the UN, effectively opening and giving credence and legitimacy to Hamas.

Formerly sophisticated people are regurgitating talking with no self-awareness, leaving people thinking there is something missing when talking with a vaccinated covid fearful friend.

Isolating people

Dr Nehls argues that the pike protein along with fear based and isolation measures are an intentional attack on our brain hippocampus. This is where individuality originate. So, with mRNA vaccines people have become less individualised, more forgetful, and more compliant. This is what we are seeing too, more conflict, less compassion and more polarised, non-thinking views, mapping neuroscience with propaganda.

Why else would the US Government have spent 100s of millions of dollars researching how to make populations more tractable, less creative, and more compliant?

Fear is the deliverable using the public health vax propaganda. Fear such as the unvaccinated are “killing your granny” creates a powerful excuse to use public health as the Trojan Horse or proxy. Fear itself too damages the brain.

Isolation creates confusion too and makes people more easily manipulated, which explains why a message that “closeness with other human beings can kill you” might rewire the human brain to create the delusion or cult like thinking that overtook the thinking of those who were normal thinkers pre-2020.

So Nehls theory is that mandated lockdowns and mRNA injections were more about creating passive, malleable citizens that public health itself.

If this is right then it explains how the cult like diktats such as Rishi Sunak “vaccines are safe and effective” whilst he has financial interests in Moderna, and Penny Mordants “Conspiracy theory” accusations. Both too pursued nonsensical measures such as isolation, masks, and submission vaccines without protest from the globalists.

This is in the context of corporate organisations and the UN increasing the narcissistic fragility of marginalised people in fabricated groups, using gender ideology and equity diversity and inclusion. People become fragile because these falsehoods leave peoples thoughts (lies) unchallenged. So, when these lies are challenged, these victims explode emotionally, unable to maintain contact and connection with people with which they disagree, because their house of cards collapses like the narcissism of a 2-year-old child.

This protection of narcissism is a great incremental tool to stop once friends coalescing around tough challenges to resolve them or rising in Revolt as part of the normal democratic process.

Its only healthy human interaction that keeps people sane and checks the validity of our thoughts, most thoughts we have are false and unchallenged thoughts become our facts.

This is how people have submitted to propaganda narratives.

So, people have been intellectually blunted physically with the spike protein and psychologically with repetitive fear-based propaganda messaging of the brain.

So, in part it's important to protect your amygdala, so avoid messaging scenes of rape and atrocities to avert this war on our brains, humanity and maintain the full range of our intelligence.

The dangers of acting like God.

People's challenge now is technological adaptation rather than genetic, and technological changes happen far faster than genetic changes. The globalists are stepping into the realms of thinking we operate outside nature, and they can manipulate our natural environment, despite having virtually any knowledge of its interconnectedness and complexity.

The key check for the globalists is governance and a healthy check of ideas, yet the level of hubris and ability to manipulate is accelerating. When previous actions are shown to have failed, so the reality may catch up with their unchecked ideas, so they have to make sure they don't stop, else like the trans ideologues, they might get found out.

Having ideas is one thing. In the past the stupid ones just died off and never got to market or got exposure. This environment of psychological corruption and manipulation of and by most governments, by the all-knowing WEF allows them to impose their plans, even though the evidence for their actions is nefarious and based upon lies and dominant mantra such as "safe and effective" or "climate change" escalating to "global warming" to "climate emergency" "climate boiling" etc, leaving these lies unchecked.

The number of seasonal variations of respiratory virus tend to mutate very fast so it's not possible to develop herd immunity so this why vaccinations will fundamentally miss the target. The secret of a virus's success is high reproduction rate and high mutation rate.

People have been so successful because we can change our lifestyles locally and adapt because of the enormous diversity of human individuality and creativity. This clearly doesn't fit with the globalists plandemic strategies because it is from this creativity comes individual ideas, some of which become enormously successful so take root so transform society for the better.

People are ultrasocial learning machines so when you isolate them, they malfunction. I noticed this with my children learning from home during the plandemic, they "didn't know if any learning was going in". In part this is because mirror neurons become active when we observe others, in this way we learn better than any other species and then pass the knowledge on. This mirroring is the source of empathy, and the fabric of the formation of self-correcting complex relationship based social systems, that steer the ship and naturally weeds out the rubbish.

The sum of the individuals x their degree of individuality = degree of innovative capacity of a society if it recognises and values people for who they are and values their difference of opinions.

The damage done by vaccines in reducing people's ability to think, has created people, driven by a raft of emotive victim centric destructive ideology. This completely trashes this ability and prevents these people from coalescing to fix problems, the perfect mess and disorder for the globalists to impose their nefarious schemes and dismiss people as increasingly inept obsolete and stupid.

This fuels the technocrats personally virtuous cycle of transhumanism melding the need for AI and monitoring because the dysfunctionality the Wef will create in this dystopia ultimately leads to escalating violence. It's necessary for them to silence the intellectuals who can make sense of what is going on.

So, the technocrats must keep peoples' capacity for individual innovation as compromised as possible to realise their technocratic dystopia, else their ideas would become a threat to the people's collective power.

In this way the technocrats can permanently deprive people of their possessions AND individuality – so that they can be happy without possessions. This however only succeeds in novels so far!

Individualisation takes place throughout our life and at birth and our brain remains malleable throughout our life. This malleability and variety of opportunity is the basis of our lifelong learning, individuality, and creativity. This variety of experience creates our unique experiences that become memory.

Our individuality is unique in relationship to others and forms our perception of the outside world: an external thing observed as actions and reactions, interests, social behaviour, and our personal views, that make that person unique.

Isolation of people from family makes people more lone fighters and dependent on the state. In schools' the government's indoctrination of what to think, sit still and obey, starts early before they can start forming their own opinions, rendering individuality absurd from the start. Learning is standardised, squandering the value of free play. Brian Sutton-Smith said, “the opposite of play is not work but depression”.

Free play creates individuality and in a business context this equated to having fun whilst working. The reduction of free play is linked to depression, helplessness, anxiety, and suicide, as well as creating a platform for ideology to be pushed increasing narcissism.

Free play without ideology or regimentation enables children to blossom, although since 1990 whilst IQ has risen creativity has reduced and kids are less able to adapt and have less mental agility to see things from different angles.

In the same way, standardisation of view when it came to Covid, closed scientific discourse, the democratic process and competition to allow the best ideas to come forwards, let alone prevail. Thus, in schools' gender indoctrination, which is based upon patent lies, succeeds before an individual can even develop their own personality or individuality so in the same way goes unquestioned.

So, what if the plandemic measures were orchestrated with the perfidious aim to rid us of our individuality, ability to think critically, and what if it was it possible for people to love that lack of freedom? Thereby technocrats are replacing the discomfort of reality with a technocratic set of lies.

Yet the absurd propaganda continues: Klaus Schwab kicked off the Davos 2024 conference with the lie that the biggest threat to civilisation was “false information”. Buy a mirror club head Klaus?

Nehls suggests that it is possible to force people to build their personalities upon false technocratic narratives, by modifying the where, when, what and how did I feel, aspects of memory. So technocratic dialogue can modify memories based upon our wealth of experience by deleting and replacing the answers held in the hippocampus and system 2 memory with uniform technocratic content. This creates simple conformity of the masses or a Great Mental Reset.

Our brain learns quite slowly based upon a lot of repetitions. But when we experience something exciting, and important, we remember it immediately and may never forget it.

Because it’s important we begin to consider the implications of that memory as if we put it into practice. But even this overwhelms our short-term memory if we put this into practice, because these short term memories are only held and prone to being lost with a simple distraction.

Out hippocampus is our store for these autobiographical memories, and it records the where, when, what and how did I feel aspects of memory.

Without a hippocampus there is no past or future, and nothing left to orientate your world around, so is a loss of humanity. You cannot develop a relationship with a personal like that. To prevent overload itself, the hippocampus makes choices what to store at the end of a long day.

The answer to this selection is how it felt and did what we experience trigger emotion or meaning. Stereotyped memories are ignored by the hippocampus because they are circumvented by the routine responses triggered from routine scenarios. So, to get someone to remember something you must package it emotionally, and **fear** is the unethical means of achieving this goal.

So, a memory that caused one to become afraid associated with it gets remembered. Short term memory and long-term memory must work together if we want to create, check, and develop new ideas.

The hippocampus must move the new information into memory storage, neocortex during deep sleep like a memory stick. These new memories are compared to existing memories when we dream during REM sleep, so we often wake up in the morning with new insights.

These memories are indexed and retrieved when the same neurons that created the original memory are triggered. Forgetting happens when the inability to create the trigger the same pattern due to lost synapses or connections.

Thus, how the hippocampus creates these index neurons dictates whether we can remember these new and old memories into old age.

The dente gyrus during REM sleep creates thousands of new neurons every day to distinguish old memories from new ones but is a bottleneck and allows a small fraction of neurons to have a significant impact on hippocampal circuitry and function.

A lack of pattern separation is compromised by anxiety and depression and linked to compromised cognitive emotional intelligence. These new neurons created are hungry for new experiences that they want to store. This is an indicator of human curiosity as well as cultural progress of humanity.

So, the index neurons make an ideal target for supplanting unquestioned narratives into the human brain, when the hippocampus neurogenesis weakens psychological resilience.

The more positive memories you have the stronger your ego or self-esteem will be, and the more confident and resilient you will be. These also protect us from excessive stress reactions. Mental resilience is a prerequisite for welcoming new experiences and for avoiding potentially stressful situations in the first place.

Stress produces stress hormones. It is better to suspect danger and be prepared for it. New index neurons are being formed every day so only they can identify and classify dangerous situations and their danger potential. But all new information converges on these new neurons by comparing to previous memories to evaluate these comparatively.

Not every new situation triggers a stress response because relative comparisons between the cues of previous recorded experiences. This is adaptive and old index neurons and new index neurons play a key role to appraise the probability and safety of the threat.

However, if the hippocampus regenesis is disturbed, if we don't have new index neurons the stress hormone rises excessively in any unfamiliar situation, shutting down curiosity and stress only returns to normal levels slowly if at all, because it's lost its indexing reference.

The fear of the unknown can create depressive illness, avoidance, and cognitive impairment. Chronic stress paralyses the production of new hippocampus neurons, creating a closed loop that is life threatening in the long run. Anti-depressants regenerate hippocampus neurogenesis.

Therefore, the production of new indexing neurons in the hippocampus is crucial for maintaining our cognitive emotional resilience but also our willingness and ability to think for ourselves.

To change one's behaviour, one must first recognise whether that change is beneficial for us or not and willingness to invest in the mental energy to think about it.

So, with a deficient hippocampus neurogenesis more and more people are afraid of new things therefore of behaving differently. Even when the change is indispensable there is a lack of will power to implement it.

Our executive decision-making part of the brain requires mental energy, which often is not available. What kind of energy does it take to imagine a new way of life, plan and implement it?

System 1 is the automatic "doing" part of the brain, that responds automatically to repetitive scenarios with no effort by default and efficiently keeps us safe without having to stop and think.

System 2 is much slower but thinks things through and is only available to us in limited quantities and when necessary, will change system 1 behaviours.

System 2 allows us to map and develop new courses of action and update system 1 as required. Depletion or ego depletion, of mental energy makes it difficult for us to think in a focused way and find the motivation to use system 2.

With complete mental exhaustion it's almost impossible to use system 2 or find the best solution to a problem.

System 2 requires the efficient storage of our thoughts driven by the hippocampus yest the hippocampus has limited storage. With a full hippocampus memory System 2 would only work if the previous memories were overwritten, but the hippocampus tries to avoid this and makes it difficult to retrieve specific memories.

In contrast system 1 or "zombie mode" is unaffected by ego depletion and hippocampus memory impairment.

An extreme example of a system 1 controlled being – leads to the progressive failure of system 2 to become increasingly reckless and tactless in dealing with others showing no regard for social norms, they uninhibitedly express their uncut thoughts in a rigid zombie like tone.

Appropriate living conditions allows the hippocampus regeneration to remain intact. The alternative is Alzheimer's.

Purpose, exercise, social life, sleep, and nutrition help prevent indoctrination.

Meaningful tasks are associated with manageable tasks and challenges. New tasks significantly increase the survival rate of hippocampus cells, which gives us a sense of a meaningful life. Taking new action in this way increases grit.

Key factors to enable a healthy mind:

Exercise stimulates brain and hippocampus cells growth and memory capacity, as well as hippocampus cell loss reversal.

Nutrition helps the mind includes Omega 3, thyroid and iodine deficiency, d3, fasting, avoiding sugar and deep-fried foods.

Social life: oxytocin is the bonding hormone. When people are isolated, this is compromised and has a detrimental effect on our hippocampus. Being with and engaging with people, creates experiences that we need for the development of our hippocampus neurons to ensure their survival.

Sleep: the hippocampus can only make neurons when it is not busy during sleep. Sleep deprivation inhibits brain growth and hippocampus growth.

Time: technology reduces the time we have to live a balanced life, which creates chronic stress. Having too much time or doing nothing is no good either because growth impulses for hippocampus regenesis are missing. Eustress, or positive stress, caused by having the time to do doable tasks is ideal. Time is an important factor in mental health.

Maintaining autobiographical memory and this the integrity of our personality depends upon regeneration of new neurons. This creates our willingness to try something new and expand our horizons when necessary or promising by using our system 2 capability to think.

The average human loses 1.4% of its hippocampus volume per year because of an imbalance. This leads to a state of mental exhaustion and lose of the ability to think for themselves even when in a rather inconspicuous supposed healthy state.

This is a clear indicator of system 2 decline and chronic ego depletion. This depletion has probably led to a population accepting freedom stealing measures including the fear, isolation, and experimental vaccination forcing the body to create toxic proteins.

Ego depletion weakens the psychological resistance of people to allow them to be controlled by irrational fears. This leads people to follow the masses or governments to fearfully follow instructions from their technocratic overlords, even when running into an economic and psychological dystopian abyss.

Numerous studies falsely accept the hippocampus decline to be a normal part of aging. It's not.

People who want to control the world by vaccines cannot be stopped by plausible science: Bill Gates conveniently showed his ignorance, when he said he didn't know why flu was seasonal. Influenzas are seasonal because of the position of the sun leads to vitamin D deficiency which weakens the immune system and causes seasonal respiratory diseases.

The reckless disregard for human life during the plandemic makes it clear that there is more at stake than just money.

Iodine and Vitamin D supplements could have reduced damage to children's brain development and preventing disease and maintaining hippocampus function.

So, the suspicion is that the real aim of vaccines and measures was to realise power, achieved by reducing people's mental capacity was the technocrat's real goal. Thereby make it possible to impose their dystopian 2030 future.

It is conceivable that the technocrats have been so successful in increasing depression by reducing hippocampus regeneration, leaving protection from indoctrination is at an all-time low so people are ready for them to **OPENLY push for the next stage of their plan without countering too much resistance.**

Distorting the decision-making process

Ivermectin and Covid being part of a Google search clearly wasn't a priority for Sundar Pichai?

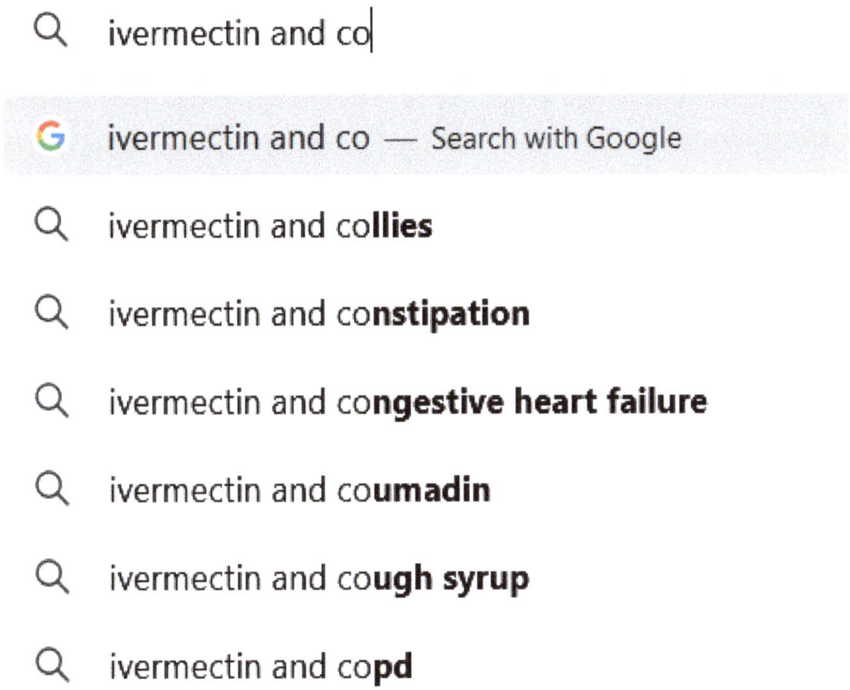

For the FDA to enact emergency legislation it had to suggest that there were no viable alternatives.

Vitamin D, C and Zinc, Ivermectin was considered a viable alternative, yet governments, ignored, suppressed, and were discredited by being labelled as horse medicine etc.

At the time of launch experts such as Professor Ian Brighthope suggested that mRMA would never stay in the deltoid muscle so mRMA presented a significant health risk to health.

Commercially there was no money in no money in Ivermectin because it is a cheap commercially available treatment. The FDA claimed that Ivermectin was dangerous, and Ivermectin proposal was ignored by US senate as well as other Governments.

Furthermore, clinical trials are funded by the drug companies. So, there was no comparison between the return on investment of a mRNA vs Ivermectin trial, the commercial decision was clear.

So, by conveniently discrediting Ivermectin, as an alternative, the FDA could enact the emergency legislation to steam roll the risks and recklessly launch the mRNA vaccine.

Yet the use of Vitamin D and Ivermectin etc could have protected the population from serious respiratory disease and enabled herd immunity without lockdowns.

This culture created a psychological infection, alongside mass manipulation. This type of challenge is a psychological infection, so not a rational situation that responds to rational debate or science based upon merit, so the Trojan Horse continued unabated:

US biotech Moderna has completed the installation of its clinical laboratories at its new R&D facility in the UK dedicated to the discovery of new mRNA vaccines and therapeutics, part of a 10-year partnership with the government agreed last year that has been estimated to be worth around £1 billion. This facility will ultimately produce 250m vaccines per year.

People develop kinship by having related memories accessible via the hippocampus neurons. This defines who we are, and without the hippocampus, we would not be able to identify soulmates and without having access to the past, there is no future to philosophise about, or any future to predict using system 2 memory, in our minds.

Without a soul there is no sense of self, no memory or anything and no chance of recovery, you just exist like an empty shell.

Our soul is the very essence of our spiritual being the hippocampus is the access to that.

The technocratic elite succeeded through Covid measures in persuading people of the narrative of the “safe and effective” vaccine, against all evidence. Detractors are met with vehemence vitriol or those that think differently at all.

The soul makes eternal life possible because the body dies at death. These narratives have a huge influence upon what we experience and how we experience it so exert control over us.

So, narratives are the attempt to indoctrinate by power hungry technocrats at the cost of considering alternatives. This is powerfully driven by our worst fears of suffering, fear of death, or fear of harming our children or parents, which underlies all fears.

Removing the cognitive dissonance from the past removes the contradiction with the technocrats' dominant dialogue and the reality of a future an AI controlled existence. The technocrats social operating system creates the counterpart to faith: so, Christianity must be destroyed.

The new social operating system is installed by wiping the old data, by overwriting the old index neurons so that people won't be concerned by alternative dystopian life plans. The hippocampus then arrives at the new future and knows nothing else, so people are happier and more satisfied than before, because there is no alternative history or indexing to alternative memories from which to alert or cause concern.

The former self is then erased so that the past would only be remembered as dark and dangerous due to nuclear, viral, and associated suffering and or causes of death…

It's a 2-pronged process that has been successful during the Covid measures:

- The production of new neurons must be completely supressed, and existing index neurons must be decimated by degeneration measures.
- The remaining indexing neurons that provide access to autobiographical memory are successively overwritten, with the indexing of the technocratic narrative.

The believers had no bandwidth for alternative narratives, such as Penny Mordant rejecting Andrew Bridgens case for excess deaths as a "conspiracy theory", or technocrats and or their puppets such Rishi Sunak repeating vaccines are "safe and effective", or "vaccination protects and restores our freedoms".

The response to a challenge to these narratives was one that felt as though the believers were being personally attacked. This feeling of being attacked demonstrated that the social operating system has been successfully installed, limited system 2 capacity to consider alternatives, or curiosity about a vital medical / safety question. evidencing the technocratic narratives had become a programmed as part of the new believer's personality, and system 2, so had to be defended.

The Covid pandemic was the first worldwide orchestrated effort to enact a mental transformation globally, impose a new set of rules for co-existence and technocratic narrative to corrupt the reality and corrupt your ability to consider alternatives or think at all.

This must be considered as a successful attack on the hippocampus so Klaus vaguely warns more, and worse crises are / will be on the way.

The Covid vaccine spike protein inhibits hippocampal growth and activates neurodegeneration, yet if this happened too quickly then the indoctrination would be pointless because people would fall straight into dementia or Alzheimer's disease, so there would not be enough neurons to overwrite.

This narrative and vaccine induced corruption of the mind, allows people to fall for a lie or staged event. The extent of the actual danger is irrelevant otherwise it would create manageable fear with far less effect.

This chronic state of anxiety for life and limb and reduction of new neuron production, makes it easy to manipulate people, so the anxiety creates a positive feedback loop that people can't escape from. This destroys of your hippocampus ability to index.

If we operate when we are exhausted, them system 2 neurons have no option, apart from being overwritten leaving the hippocampus unable to access the old memories.

Fear happens when we face an immediate threat, anxiety is the anticipation of a threat. Fear drives system 1, such as fight or flight, when anxious best to kick in system 2.

When people are held in a fearful state narrative gain access to overwrite system 2 memories, by delivering the narrative in the state of ego depletion and with sufficient fear.

If people are faced with access to anxiety inducing news when the ego is depleted the technocratic narrative activates system 2 and replaces the system 2 narrative with the technocratic narrative. Without unused neurons in the hippocampus autobiographical memory access is removed so access to these memories is erased.

As a result of the depletion of self and ego, alongside fuelling narcissism weak people are being created, where dissenters are pests and are only comfortable with mass conformity.

Confusing contradicting, absurd media is key to maintain mass psychosis and maintain panic. A constant repetition of the same would be ignored because that had already been rejected by being cross referenced to previous memories that ignored it.

Thus, with more and more varied nonsense, more of the old identity is replaced with a new self-image driven by the technocratic narrative.

These weak individuals are ideal candidates to use hate speech and taking offense as the means to fuel the escalating division of society along with using the baseless virtuous ideology to isolate and cancel individuals to deprive these people of healthy social function that keep their minds healthy. So, the wrong people thrive.

This varied nonsense maintains the type of fear that does not stack up with reality. The fear is not concrete, thereby obfuscating the reality, and causing people to question their own sanity. In this absurd confusion a weird technocratic narrative in such an absurd scenario, provides comfort and purpose, hence people take the easy path by religiously following the gender, climate, vaccine nonsense etc, whilst idolising and associating themselves with the achievements and wealth of the technocratic elites.

The crux of creating anxiety and fuelling a mass psychosis is: having a flexible timeline, not too short term or people will figure out it's a lie, propose potential solutions, creating a consensus using scientists that agree don't worry about lies, (most people don't fact check), using dishonest scientists / academics / media that will accept funds to back your claims. Then switch to a new absurd scenario before people figure out that the previous one is nonsense, then repeat the same steps…!

This is why the technocrats want to move quickly to impose their dystopian tech agenda so people can't keep up and if they slow down, they will be found out.

Sadly, along this mind-bending journey people's mental cogency is diminishing and their ego being usurped by the technocratic narrative.

This technique has been known since the 1950s and it seems that this has been applied with scientific precision to the entire global population for the first time in 2000.

Fear eats your soul. The Gates Foundation funds research into PTSD which is shrinks the hippocampus. Bill Gates led the Covid 19 anxiety campaign making senseless recommendations such as lockdowns, injections of viral genetic material, and more, to “save humanity from the virus”.

BUT. The reality was it was not the virus that kills but the overreaction of the immune system due to lack of Vitamin D hormone.

Stress pre and post stress.

There is little difference between prestress and post stress disorder, prestress disorder was the key tool to scare the population during the Covid Crisis and keep the people in an anxious state through a wave of nonsensical and conflicting actions.

It's important to note that politicians were vulnerable to these phenomena too, by obediently abdicating or leaving the puppet master technocrats to pull their strings. So the politicians became a key tool to maintain the state of mass psychosis with every weirder and conflicting solutions wave after wave, before people have had a chance to check them out.

The politicians obediently publishing people's worst fears by sharing the daily deaths, creating the most fundamental anxiety that that death might happen to me or mine etc.

It's hardly surprising that the World Economic Forum launched their Global Risks Report of 2020 in 2020, each risk being interrelated and amplifying others. This optimally served as a pre traumatic stress report for the general population.

Climate change has created acute mental health effects, more than half of Americans believe that global warming is primarily caused by humans even though humans only produce 4% of the Global Co2. It is another one-sided discussion with all the same nonsensical horror aspects, that lead to mass psychosis, just like Covid 19, was used to preserve and excessive expansion of power to fuel the Great Mental Reset.

It is enough to make everyone **feel** like the majority feel that way to sway the masses. This creates a self-fulfilling prophecy and a blind spot.

This uses the social compulsion "to belong", that a majority is formed despite and regardless of the facts or the reality in front of people's eyes.

Fear leads to bad and irrational decisions, such as being spiked. For example, Covid 19 mortality rate 0.35% which was hardly any higher than the normal flu rate and continued to fall.

If adequate vitamin D supplements would have been used this could have been reduced to zero, although Bill Gates, the dominant media force had no idea why flu infection rates are so much higher in the winter. Vitamin D was the causal prevention that the pharma companies didn't want.

It is also clear that vaccines, given the ever-evolving nature of virus will never be able to cure or stay ahead of the variants and mutations.

Positive generalisations are as dangerous and as effective at creating blind spots: the statement by Fauci "the unprecedented speed with which safe and effective Covid 19 vaccines were developed, proven effective and distributed in millions saved lived" made no mention nor left no space for ineffectiveness and serious side effects = the blind spot.

This leaves Moderna developing mRNA vaccines for many other diseases yet spiking never proved effective against infection.

Given the reality it is far more likely that the vaccine was intended to destroy the autobiographical memory as a precursor to effective indoctrination to enable the dystopian Great Reset agenda. . Those that have received multiple injections must instead now live in anxiety for a whole range of late effects.

Who found out that mRNA was a bioweapon?

In 2007 Chinese researchers discovered in Wuhan that the spike protein of Covid 19 created large amounts of inflammation messengers, so attacks the natural function of our autobiographical memory via our own immune system.

This assault also operates across the placenta and impaired spatial learning. Prolonged inflammation reduces hippocampus volume and contributes to neurodegeneration. Lipid nanoparticles enable these mRNA or DNA particles to penetrate the blood brain barrier. So, by using these Lipid nanoparticles it suggests that they were intended to penetrate the blood brain barrier.

Moderma found that 2-4% of the mRNA was found in the brain despite having been injected into muscle as usual. The packaging itself produces inflammation too.

The revised WHO definition of a pandemic does not relate to a threat to life, just by having no immunity to a threat the, WHO can conveniently declare a pandemic. So, any genetic variant no matter how harmless would-be grounds of declaring an emergency.

Open and honest conversation would have prevented people being injected with these brain damaging substances.

Bill Gates despite his conflicts of interest being invested in vaccine research, long before 2020 was secretly the head of the WHO and initiator of the GVAP strategic immunisation goals setting out how the world community should behave. Vaccinations were presented as a weapon with no alternative, and warned of a pandemic horror scenario, difficult to distinguish from an advertising campaign, or a war that doesn't include weapons rather a highly infectious virus.

The pharma contracts hold the manufacturers harmless and payment of ordered vaccines is guaranteed, even if a better product exists and there is no right of return.

These terms fuelled compulsory vaccination by the state and a degree of duress despite the vaccine didn't work from the beginning. No protective effect was expected anyway, due to the constant state of new variants. So, gag contracts were the only means for governments to save face.

In Gates opinion "bioterrorism is the next big threat" and "we don't worry enough" the vaccines themselves are a serious attack of this kind.

Unfortunately, Gates predictions have been miraculously accurate, so one must take them very seriously!

It seems clear that we are involved in a multistage process and Covid 19 was only one phase and protecting health is only a pretext.

As Klaus Schwab's book is entitled "COVID 19 – THE GREAT RESET, how true its clear how financial interests are not the only motive, the Great Mental Reset is in progress so power, disabling the neuro capabilities of people are perfect strategy to enable the 2030 agenda.

The real solution Vitamin D that could have solved Covid 19 from the outset was absurdly labelled as" right-wing extremism".

Power and stupidity – the winning formula

Power is defined by the degree of subjugation over those, by those who claim to have power.

So, it's the ability to control their brains and manipulate people at will. Bonhoeffer said "The person manipulated becomes stupid, but this is not because he lacks intellect. This stupidity is a more serious enemy of the good rather than the evil: evil can be countered; we are defenceless against stupidity.

No amount of reason helps and arguments fall on deaf ears, facts that contradicts a stupid person's prejudice simply need not be believed, and if there are refutable, they are brushed aside as unimportant or a conspiracy theory, a derogatory claim that the suggestion is infantile.

An effective way to produce this type of stupidity is to switch off neurogenesis of the hippocampus, the spiking fear and anxiety induced even though the real threat have been minimal, so far!

The crus of governments: the strengthening of the economy was abandoned at great speed when the lockdowns were imposed, throwing that most basic paradigm that drives our social and economic wellbeing away, by escalating anxiety of catching the virus.

Loneliness has striking harm to the brain correlated with shrinking the brain and reduced cognitive processing weaker emotional regulation. Plandemic anxiety increased from 6% - 51%, depression from 15%-48%, PSD 7%-54% stress from 7%-82% in the general population in China.

Long term exposure to mobile phone radiation damages the hippocampus, yet the roll out of the far more intense 5G is being rolled out like society depended upon it.

Children are more vulnerable to this radiation, so digitising our society is a risk.

Devra Davis suggests if cell phones were a drug we would have banned them years ago, 5G would have never been allowed to market."

Monotony has a detrimental effect on the hippocampus, for example hippocampus volume decreased by 7.2% during 14 months of artic isolation even though the explorers travelled in a group.

Wearing facemasks and speaking on zoom leads to loss of verbal and non-verbal communication cues which makes it harder to identify emotions and masks reduced oxygen.

The use of vitamin D would have ruled out emergency use of the Covid vaccines from the outset. Scientifically published papers showed that serious adverse effects were far higher in those that took the vaccines from Pfizer: 52% - Johnson and Johnson: 323%.

So mass vaccination is hurting the health of the population in general. Vitamin D also reduces anxiety and anxiety symptoms, but discrediting vitamin D is instrumental to the "Great Mental Reset".

Given this immense damage, the population will need help at an early age or even die prematurely.

In Huxley's world the elderly play no role for the world controllers anyway. So, if one wants to establish a new technocratically run society it depends far more upon children to young adults who have been exposed to technocratic woke narratives pushed in their technological and corrupted educational bubbles. So, these children are unable to resist susceptibility to these narratives with anything of their own, because they have already been programmed on a day-to-day basis.

The fertile ground of this weakening was created by vaccines and the various other lockdown measures and technocratic indoctrination during the children's education which discourages free thinking or thinking for oneself.

Focus on corrupting children.

The ideologues focus on education because, children are particularly susceptible to narratives, because they haven't developed their own narratives, so they blindly accept narratives as useful patterns of interpretation if it is told by a person a teacher or parent that they hold in a position of trust.

Imposing new narratives on adults is more difficult you must usurp a world view and their common sense.

With adults the more absurd the narrative to be internalised and then can be **enacted**, such as forced / mandated use of pronouns, the greater the power is exercised over a person. This is the goal of triumphing over common sense. This is coupled with the human need to "fit in" or the extreme discomfort of being cancelled, people tend to follow the masses regardless of the facts.

People also rarely checking the facts. So, math is now racist, 2+2=5, a woman is a man, pronouns, ESG, gender ideology, equity diversity inclusion, climate emergency, vaccines etc you now must believe it and enact it…

The more absurd the better. So, the power of the technocrats is greater than the reality of the threat.

The more absurd and meaningless the narrative accepted by the broad masses, the more monstrous is the control of the human brain. This is why the UN want to legislate against and criminalise any detractors of LGBT ideology.

Decoupling the reality from the technocratic imposed narrative from the human brain, makes reality no longer a check and balance. Covid regulations were never about health but about power.

Eradicating differences between people such as sex and culture, is eradicating what makes us human beings. “Whoever is no longer capable of thinking differences for him they simply no longer exist, no matter how openly they exist.” Nyder.

The more we try and resolve this nonsense with our rational minds, the more efficiently they become part of our autobiographical memory this part of our personalities.

Contradictions are remarkable for the hippocampus because they are contradictory masks and good, masks are bad etc. So, because of this disturbed indexing and neurogenesis, indexed neurons are increasingly overwritten, making the earlier experiences inaccessible = erased.

This is dangerous because our past memory, is our reference to see and means to make sense of the world and erases our ability to visualise the future. So, our personal reality is distorted by the stories our brains tell us.

The contradictions also provided a way out for politicians and a sense that they were just doing their best, and that they appeared not to know what they were doing. This perfectly obscured and orchestrated the top-level plan and obfuscated any responsibility of the technocrats.

This governmental stupidity justified the existence of the technocrats, so they claim everything must be coordinated more closely and centrally.

The WHO site the narrative of “catastrophic Covid 19 failure and only a WHO can fix to prevent worse things happening in the future”, by subordinating their own interests to the interests of the community, in essence making the community the victim of idiots.

Yet the reality is that the technocrats have become obscenely wealthy at the expense of the community, conveniently without a clear post reset plan... just platitudes and fear and no accountability.

The grave concerns.

The grave concern is that a large part of the population is neurologically damaged so no longer able to resist any stupid technocratic narrative already, no matter how monstrous or absurd it may be compared to 2019, the spiking, permanent crises and the covid measures.

The North Korea and Chinese brainwashing include getting people to admit they have falsely done something wrong and confess it, alongside vague threats which create pre traumatic stress disorder to induce compliance.

The capitalist market economy for stakeholders is being corrupted to be replaced by an economic system to suit the technocrats. This system isn't communist because the stakeholders are held in a state of fear and anxiety the apocalyptic horsemen of the WEF.

So, "the ordinary stakeholder needs to be protected from himself, property, free of human rights, legal rights, and completely supervised by the WEF and Gates. This is the only way to save us from the imminent end of the world".

We only must let ourselves believe that a vaccine and climate neutral world is only possible by taking the next step by becoming transhuman, getting chipped, CBDC, Ai controlled and sacrifice any power we have over ourselves. Then and only then is there a chance of continuation of human life and we are provided with what we need to be happy. Spiking is useless for immunity its great for autobiographical memory destruction, so take your jabs every 6 months.

Those that don't conform get smugly treated like heretics. Criminalising people for not conforming or falling foul of hate speech laws, or not taking vaccines caused in 2023 a German health care worker to be accused of the involuntary manslaughter of a patient for not taking her vaccine, even though her taking the vaccine would have afforded no protection for the patient. People have also received visits from the UK police for speaking out about Covid measures.

Sowing discord divides and rules: destruction of the topic of debate creates isolation. So controversial gender or race topics have been introduced into the discourse alongside their radicalised advocates.

This leads to rifts in friendships, and families. Battle lines get hardened, and people increasingly think of extremes such as misogynists, evil doers and advocating for third-party victims.

Opponents of gender speech are radicalising themselves as well and see themselves as saviours of moral code or protectors of language, or on the other side, as seeing gender advocates as paedophiles.

This makes these groups irreconcilably opposed to each other, in a binary way, with no graduation of argument, so one can be lonely among people leaving them with nothing to say anymore.

People are an indivisible social element that cannot exist in isolation. Nehls argues that the nuclear family is the “molecular structure to keep the atoms in a safe place”.

To break these social bonds instead of molecules of requiring energy breaking social bonds requires anger, hatred, and division. This makes the person with a different view an isolated person vulnerable and attackable.

Gendering or vaccine status etc is part of this, by falsely defining whether a person is a good person all which people must adopt to fit in.

This divisive narrative creates the space for the technocrat’s protagonists to fuel the fire and divide people, without getting their hands dirty and drive their agenda, whilst the population is unable to coalesce.

Fuelling narcissi

Biderman notices that Isolation and monotonous routines cause people to become obsessed with self, so this fuels narcists. Combined this with stress and anxiety and you have depression as well as being conveniently completely detached to those technocrats responsible for this dilemma.

American soldiers captured by the N Koreans who defended Communism and the Swedish captors refused to testify in court against captors and even raised money to defend their captors. Supporting one's hostage taker is a coping mechanism to cope in an abusive situation and isolation creates a confidential relationship between captor and hostage taker. This is known as Stockholm Syndrome.

Most people realised what was being done to them but very few have raised lawsuit, because the tyrannical confusing dialogue is maintained "we are vaccinating people back to freedom" said Spahn a member of the WEFs Global leaders' program. So, the captor can only become free by obeying the hostage taker, whilst the technocrat's lie by stating the trap that it the vaccine and measures "are only for their own protection". This damage done by the vaccines could easily create a dependence upon the WEF hostage taker or entrapped into a life of permanent dependence upon healthcare.

This division and confusing dialogue provide the technocrats, via their puppet ministers, which by now are complicit too, so part of the Stockholm mess too, to obfuscate the truth.

The CDC in July 2021 study realised that the vaccination program did not reduce the rate of infection, contrary to political claims. So, the technocrats' politicians must have been driven by other motives or maybe the ministers have become puppets, like rabbits in the WHO/WEF headlights by falling for the technocrat's narrative. Let's not forget that the ministers have been living in the front seat of the technocratic vaccine narrative and under intense pressure and fear under the covid measures too.

In Germany 2022 mask restrictions were lifted for those who had been vaccinated, giving the vaccinated people the feeling that they were not contagious, promoting infection and the division of society.

Spahn's autobiography title was "We will have to forgive each other a lot" fuelling the division and by no means a statement of equals amongst equals. The Stockholm syndrome of the victim assumes that the perpetrator and victim share the same values: so, an apology is still missing.

Pushing fearful narratives from Ukraine or Covid upon a depleted hippocampus just before bedtime with coffins, weeping relatives etc overwrites index neurons with fear inducing stimuli ideal, perfect to indoctrinate.

Joseph Stalin who dies in 1953 noted "the death of one man is a tragedy the death of a million is just a statistic".

Limiting varied experience

To place an entire population completely under the control of a narrative requires limiting opportunities for new experiences, by closing pubs, sports facilities and church is key to maintain control. Communist education exposing people to nothing apart from Communist propaganda, their view of history and events.

Thus, lockdowns and vaccine strategy for the first time ever mistreated their citizens with the same rules and slogans, with the narrative controlled by mainstream media. Defamation and cancelling using the victim, scapegoat and terrorising method, thereby isolating detractors for attack, or those with alternative views, in case they and their truth became mainstream. Critics were slandered and called covid deniers, conspiracy theorists etc, escalating to right wing radicals or Nazis.

The power of this grip corrupts institutions. The FDA remained silent until a court made a ruling against them. Pfizer was claimed by Brook Jackson to have reported to the BMJ, that Pfizer had falsified and corrupted the vaccine trial leaving the Pfizer quality control people overwhelmed with the volume of problems they encountered.

Anyone that spoke out against compulsory vaccine was labelled a right-wing Nazi. From Mengele's Nazi experiments on Jews, the Nuremberg Code was formed based upon not allowing experiments or treatments on people without their consent.

The people that come up with these corrupting, divisive narratives are extremely adept at manipulating public opinion, because even high-ranking respected scientists got the same treatment and were silenced.

Yet Gates suggested there was only one way to fix the problem and interpret events, leading the normalising of the dialogue that "all the experts agree". This is patently false because the critics were just slandered and silenced. Academics have the greater propensity to be duped because of their majority have left wing tendencies.

Existing protocol such as the Great Barrington Declaration that was created to protect the vulnerable groups and let the rest carry on with their lives. But this did not suit Fauci's narrative.

The consequences of silencing are devastating, heavy censorship of critics that countered the political narrative, was carried out by Google Twitter and Facebook with the support of governments. This suppression damaged the reputation of dissenting experts and the medical profession overall.

This has led to the unquestioning of end-of-life protocols to COVID-19 patients in UK care homes, when they were still active and only mildly ill — including the use of drugs like midazolam + morphine that may be linked to patient deaths.

Each person reported that their elderly relative was administered an end-of-life protocol when they were still active and only mildly ill. They all subsequently died.

Why are discredited bodies now right?

As Idi Amin said, “There is freedom of speech, but I cannot guarantee freedom after speech”.

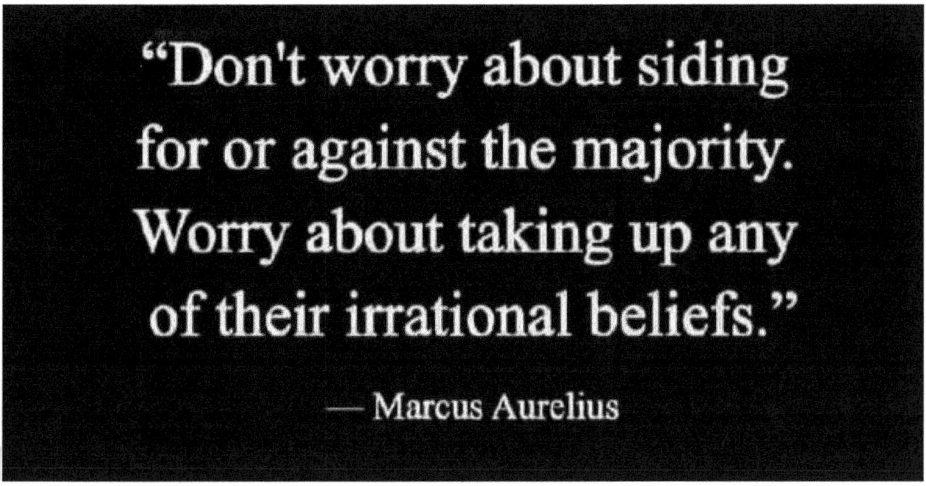

Framing dissent such as right-wing extremism, or Nazis is a skilful and sociopsychological tool to influence the way the recipient receives or rejects the narrative.

This false framing causes the hippocampus to automatically reject the counter narrative without reason or thought.

For example, the AFD was the only political party that challenged all of the covid measures with rational medically sound suggestions, such as using vitamin D etc which would have saved many people, but it was rejected by the votes of the entirety of the German parliamentary groups. This was rejected and legitimised by framing all alternatives put forwards by the AFD by being right wing.

This was easy to accomplish given the AFDs leanings, yet the challenge made no legitimate counter arguments to the AFDs sound policy. The same happened in Austria with the FPO because of the FPO party’s political classification and orientation, so if people are categorised unfavourably debate has no chance.

So, the issue is a psychological manipulative challenge to resolve, not a rational one. Why is that the solution can only be found in discredited bodies it's because they care about national interests and have been slandered.

The AFD take a position on climate change, privacy, social scoring and CBDC. These policies are all part of the technocratic takeover that are being promoted by the other German political parties.

But it should NOT matter who raises that point, if it's valid?

This shows the counterintuitive power achieved by ideological equity diversity and inclusion types, whose left-wing tendencies lean them psychologically towards entitlement, government control and victimhood.

It's a take-over by and revenge of the nerds.

On one hand the technocratic media pushes narratives and solutions that are incongruent with the reality on the ground. Thus, depriving, and negating people of their value coming from their individual experience and view of climate change, covid measures, Ukraine, or Israel etc.

Weaponised speech

You can observe from the weaponised speech coming from university protests relating Palestine, the contagion aspect in this polarised environment.

This is because this framing creates a very comfortable position to hide behind, grandstand and virtue signal in a way that is devoid of responsibility or accountability. This ideology is cosily fuelled by the lecturers who too are now 90% left wing leaning.

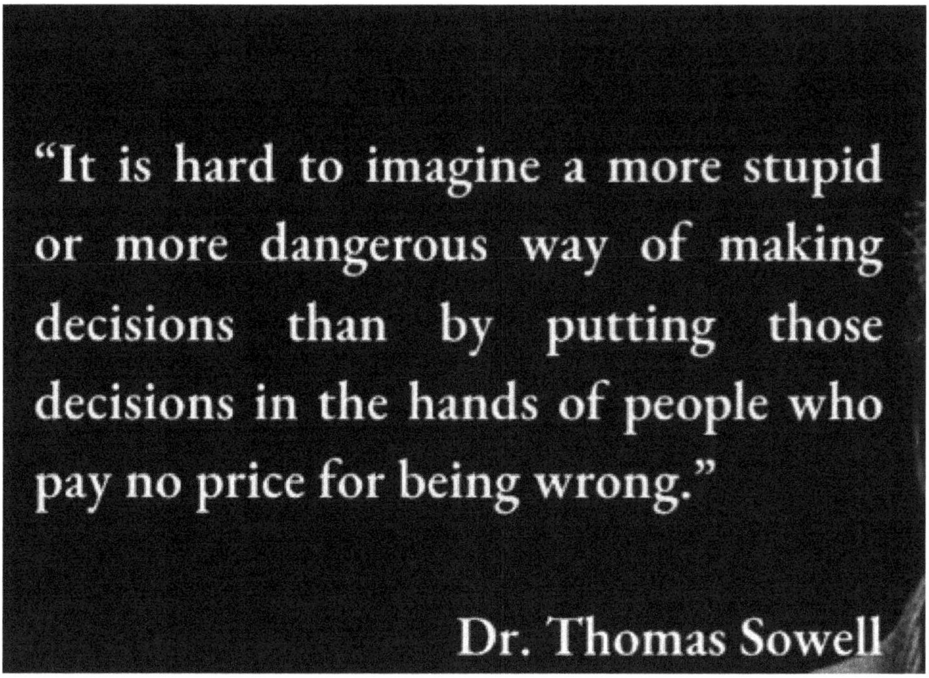

The mechanism of victimhood overwhelms the reality in an absurd way resulting in the destructive "Queers for Palestine" narrative, when pushing that narrative is devoid of the dangerous reality and consequences of being a queer in Palestine but is justified in the ideologies mind because fulfils the common virtue signalling, untouchable victim-based psychology.

By controlling the dialogue, the technocrats have falsely positioned the Palestinians as victims, when it was Israel who were attacked on Oct 7, or they try and draw broader historical weaponised negative generalisations to demonise the Israelis, when surely they are the most victimised and murdered race on earth?

The media monopolises this framing and classification as well as falling for their own narrative without realising that they have been indoctrinated too.

This is rolled out through its subset of tyrannical co-opted institutions and even the BBC that fall in line.

The media adopt the position of the Stockholm kidnapper by using narrative to manage anxiety. Yet people are dependent upon their source of media information. This keeps people in a continuous anxious state as well as offering a vitriolic sense purpose in the sea of lies for the nerds to join and defend without any accountability. Countering this predominant one sided / lies based narrative is usually met with violence.

Escalating dialogue

A threat of death triggers primal anxiety cells in the hippocampus, causing people to take a course of action that promises a way out without thinking.

This explains a whole range of irrational behaviours and strategies such as largely meaningless PCR testing or Co2 reduction and climate emergency etc.

In the past being excluded from one's group meant a death sentence, which explains this is a primal anxiety and motivation to conform and fit in.

The unvaccinated were even blamed, so victimised, scapegoated, and terrorised. In 2017 Frank Mongomery said, "we are currently experiencing the tyranny of the unvaccinated" This dialogue weaponises so escalates people in an already fearful state. This fuels further division and anger of vaccinated people, encouraging them to blindly to turn on their compatriots. This creates a social addiction to vaccination, where people have no control over their actions the same is true with the escalating existential dialogue relating climate emergency, Palestine etc...

Orders for vaccines were far more than the two each that had been planned. Obedience was achieved using scapegoating so that prison might be the obvious next step for non-conformity.

Social isolation is also detrimental to health, so we have a choice between two evils, this is the perfect emotive set of positions to manipulate to the technocrat's heretical thoughts and advantage.

Whoever goes against the WHO is a right-wing enemy of the people.

Relating the Islamic threat to democracy in the UK Rishi Sunak softened the blow in his recent speech by talking about the danger of Islam in the context of the threat of Islam and the UK's right wing, thereby rejecting and marginalising the validity of the dissenters of Islam, who specifically and intentionally undermine UK institutions with their own Islamic ideology and political interests such as Sharia law etc.

Karl Lauterbach Spahn's successor said, "In Germany it's not enough just to annoy the unvaccinated you have to do more", so food wasn't distributed to the needy unvaccinated.

This is not far from deporting or executing the unvaccinated. It is the same logic that justifies shooting down an out-of-control aircraft, to save the thousands of city dwellers that it's about to crash into.

Scholz in Dec 2021 said that "as far as he knows the infection that affects us all comes from the unvaccinated". The power of weaponised language to fuel the victimhood, scapegoat and terror using a negative generalisation.

The largely senseless measures imposed on our economies has left many dependent upon government measures. This has made them far more easily manipulated because with the various escalating threats, the message taken was that resistance was futile.

The unvaccinated were framed as being unintelligent and untrustworthy but the opposite wasn't true, the unvaccinated did not show such discrimination.

Boiling the frog

Once compliant, indoctrination is a complex step by step process to get people to comply to and action things that bear no sense to reality.

It's a slippery slope, because one action is performed that paves the wave for the next. Put it another way if you bend someone out of shape on one dimension, such as forcing compliance with pro nouns, or believing black is white, a woman is a man etc you bend that person our of shape on all dimensions. Step by step, people tragically fall into complicity and a quagmire.

The salvation of the vaccination then depends upon the narrative: masks, isolation separation etc. So successive waves of nonsense and anxiety caused these nonsensical actions to be accepted reducing the will to resist with each step. Each action became a habit centred around victimhood and the virtue signalling narrative. These steps were piled high until you had accepted the complete/ larger package of measure that would never have been accepted from the beginning, by spooking the audience.

In the case of the vaccine there was no mention of the nature of its salvation, that went unquestioned: when in fact it was systemic injection of genetically modified biologically active viral material!

Step by step on wider fronts we followed this propaganda, so gave up our freedom of speech and privacy, for the sake of fitting in and avoiding the conflict with our new oppressors our family and neighbours.

The small part of a population that can still express themselves in public who wanted to suggest the money trail to a whole load of cash, is a criminal. So, these changes are nudged and legislated for later. Hospitals became places of infection so the fear of being infected by going to hospital, became more of a concern than dying of a real disease, so there are still people wearing masks.

The apocalyptic WEF report Trojan Horse narrative continued regardless, suggesting in 2023 "global complacency seems to be setting in on preparing for future pandemics and other global threats".

The ability of people to speak out and have choice may change as the technocrat's scope expands denying access to servers, impose a social scoring or CBDC system.

The report "Covid 19 the case for social scoring systems was published on Sept 15, 2020" to make it easier and more rewarding to comply with public health guidelines, the favoured option is the Chinese social scoring model, maybe climate change will be used to leverage that narrative, well beyond and with less need for more indoctrination, because people are already weakened and compliant?

CBDC is positioned as voluntary, because no one can escape it, so the voluntary aspect is not reality just window dressing. The Prognos study suggested that the driving force behind social scoring was climate change, punishing people for their environmental footprint, without regard for whether people are responsible for it, which they are not.

Scientists Struggle to Understand Why Antarctica Hasn't Warmed for Over 70 Years Despite Rise in CO2

BY **CHRIS MORRISON** 29 JANUARY 2023 7:00 AM

History shows that only a small minority will question the points system, or digital nervous system, so we should make sure that these metrics these should be publicised, and implications made clear.

The digital nervous system closes the gap between values and behaviour by imposing the most tyrannical ethical/religious system that will rely on chimping: chipping and vaccination in 1 go. CBDC allows not only behaviour to be controlled but also money, so a dissenters' money is turned off.

The Smart City Charta post ownership society, "I own nothing, have no privacy and life has never been better", so people that don't comply will be insolvent and have no future, and find it almost impossible to get reconnected.

We have a chance to stop this if we think for ourselves: although, Bertrand Russel said.

> *"Most people would rather die than think for themselves".*

The technocrats may fall for their own narrative too.

It seems that governments have already fallen, and it may be that ministers are most vulnerable to this type of corruption. Ministers have such huge responsibility, and so could have no or little awareness of how they are being manipulated and or what is driving them.

Technocrats seem now to be believing their own spell themselves so are driving a new wave of actions to establish their worldview: that they know what is best, with their victim and punish centric and virtuous positioning.

This is leading to WEF statements such as

- The United Nations General Assembly has declared the ability to live in "a clean, healthy and sustainable environment" a human right.
- Yuval Harare: "humans have no soul", are just animals that will be hacked with vaccines.

Yuval Harare in his 2017 article: The rise of the Useless Class said, “what should we do with all of the superfluous people once we have highly intelligent nonconscious algorithms that can almost everything better than humans”.

Albert Einstein said “I fear the day that technology will surpass human interaction. The world will have a generation full of idiots.”

And who will need idiots?

How many people does AI need... the WEF calls for an 86% population reduction.

Yuval Harare answered “the best we can do at the moment is to keep them happy with drugs and computer games”: A means to help subjects reconcile themselves to bondage as their fate.

Political correctness in AI would enable it to protect a certain group of people at the cost of the rest: its creator or its programmers.

"The short-term impact of AI is who controls it the long-term impact is whether Ai can be controlled at all" Stephen Hawking.

Climate emergency and useless humans

From the technocrat's narrative it seems that man-made climate change has no other cause except man-made C02. Yet humans only produce 4% of the total global C02 and it is currently only 0.04% of the atmosphere and Co2 has been 5 times as high at .2% and the USA is cooling.

Co2 is the ultimate control of the population because co2 is probably the best means to entirely track to the economic success of the global population. Yet what is co2 it is plant food.

Bill Gates said in his Ted Talk 2018 "The question has for factors," "so you have one thing left co2, that you want to get to zero and that's going to be based on the number of people" Now is we do a really great job on new vaccines health care reproductive health services we could lower that by 10-15%.

So, are the excess deaths due to spiking part of these excess deaths? WHO publications from 1992 show a long-range purpose "to reduce population in unstable less developed countries", so in 1994 the multiple hCG laden tetanus toxoid injections were only for women of reproductive age, which was odd because tetanus injections normally last 10 years.

Mosquitos could carry vaccines, although that's illegal because that's administering vaccines without consent. In 2008 Bill and Melina Gates provided cash injection for flying syringes at Jichi medical University.

An article in the guardian of 2017, suggested having offspring is climate killer number one, and having one less child is the greatest way to combat climate change, because a child emits 58.T of co2 more harmful than 24 cars at 2.4T of co2 per car.

Christian network for Europe suggested 40% of 10,000 of young people between 16-24 now did not want to have kids because of climate change. Some renounce children at all for the sake of the climate.

But of course, every useless human emits CO_2, which is where the climate technocrats and people come into a deadly conflict. So, must we wait for the climate problem be solved for us or used against us?

We must get ahead of this manipulation and develop our own strategies to get ahead of this, after all democratic governments are our servants not the other way around.

Getting a grip and the point of no return

"A new scientific truth doesn't triumph by convincing its opponents, its because its opponents eventually die, and that a new generation grows up familiar with it" Max Planck

Because scientists make dogma famous, and are famous for their dogma, they will defend it with all their might for the rest of their lives according to a study that posthumously confirmed Plancks assumption.

Its clear that the Global Mental Reset or brain damage underpinning the Great Reset is the platform to dupe the masses.

So, we must upskill and strengthen psychologically to push back this tide. We don't have the luxury of time to get ahead because there is a point of no return.

Nehls suggests that spiking not only blocks the capacity and maintenance of our memory system that makes up our soul/ego, but also has a neurotoxic effect by making our indexing neurons the target of the vaccine killer cells. If this is the case, then for the technocrats' vaccines have been a huge success in enabling their GREAT RESET comprising of completely undesirable dystopian objectives. The spike in no way justifies protecting against a respiratory virus or vaccination of children, for which the virus presents no risk in the first place.

Why mass vaccination vs those at risk?

The virus only ever presented a small risk to the elderly, so should not have been applied to the rest of the population where herd immunity and vitamin D etc would have sufficed.

So why was the whole population applied with the vaccine and the covid measures? It does not make sense as the spike doesn't fix the vaccine, so the impact known before launch: to indoctrinate the technocrat's narrative into the whole population and overwrite one's memory is aligned with reality.

Indoctrination enables individuality to be lost. This is our central source of strength and identity and means of assessing the actions we take.

The failure of this policy to protect against the virus weirdly is being used as the basis to justify strengthening the WHOs control of society next time around.

In the meantime, the bold 15-minute cities low emission zones, completely devoid of justification or analysis, are being rolled out, and CDBCs being developed by the G7 nations. Legislation is being rolled out to reduce privacy protect the technocrats from liability.

It may be true, that with the escalating division and anxiety people have for the future and the immediate fear of disease x and vaccines, that the masses can no longer mobilise themselves to resist.

So, like any addiction, where one has lost control of their actions, where people don't tend to do their own analysis, legislative actions are required to hold the technocrats, like drug dealers, to account. We must upskill and strengthen to resist further ingress.

It maybe that the WEF and other technocrats and governments are the most indoctrinated so will take this victim centric false ideology to a whole new tyrannical level, alongside AI and intense surveillance.

This intrusive anti privacy transparency has caused little outrage, which is hardly surprising with the myriad of malevolence taking place globally justifying us human and as part of humanity to hold our heads in shame.

Courts during the pandemic typically upheld the technocrats' positions and continue to persecute the critics such as Trump.

At present everything seems to be going to plan, Klaus Schwab's recent book was even called Covid19 – THE Great RESET!

If they stand to completely usurp global power, they will do so without trial.

He that is silent is consenting, he is not resisting is accepting what is happening to him.

Claude Junker said “we decide on something, we leave it lying around and wait to see what happens. If no one kicks up a fuss because most people don’t understand what has been decided, we continue step by step until there is no turning back” Nov 30, 2019.

The sum of the current wars is just a war against our brains.

Maybe after the collapse of our economies, due the climate emergency strategy and take over by China etc we will be ready for owning nothing, have no privacy and being happy?

Why do the technocrats want to bring disaster to this world?

Bill Gates stands out as being the potential mastermind and leader. People are already talking about the “Gates of Hell”, preaching health and happiness through pharmaceuticals, and throwing everything human and of faith overboard.

These people have the financial resources and political influence and business model to act accordingly. These people are now not afraid to create viruses with new pathogenic properties furin cleavage sites or discussing this in advance in public.

They are motivated therefore to reduce human mental and physical health to reduce the resistance to their plan, with a cold-blooded indifference to the impact on people’s lives and democracy, to achieve their own goals.

Mental destruction is accepted with approval and probably a prerequisite for success. A dramatic increase in mental health and capacity and IQ in society, since 2012 is evidence for the impact of the plandemic.

Munchausen, aka the Baron of Lies, is a similar wreaker of havoc, who invented illnesses to demand medical treatment, claiming to be a self-sacrificing hero to save humanity in his own way.

It’s clear that it’s the super-rich that are driving environmental destruction with the mining and processing pollution associated with automotive electric battery manufacture, deforestation, and flex crops.

These industries have vastly increased the wealth of the HNWI. So, it’s the ordinary people, not themselves, that cause the issue, whilst they save us from our own destruction that falsely justifies their Great Reset.

Vast intelligence and power of the elites?

Maybe their vast wealth indicates their superior intelligence, alongside the dangers of maintaining and protecting power as discussed in the previous chapter.

A Swedish study showed that the super-rich top 1% were in fact less intelligent and scored less well on cognitive ability than the group below.

Keltner's study found that powerful people are more selfish, ruthless, and narcissistic so receive less advice and mirror less well.

So "powerful people stop simulating the experience of others" creating an empathy deficit.

Henry Adams suggested "Power changes the brain, killing the victims' sympathies" "acting like someone with brain damage".

This may explain why they bring so much disaster to the world, and there is a common attitude of the powerful to disregard the powerless, because power leads to a lack of empathy and leads to negative attitude towards others.

If you are powerful, you are more likely to think that you are doing all the work, others are lazy and unreliable. So, others need supervision, censoring, told what to do and monitored. So, you feel that this monitoring should be entrusted to you. This may be a God type complex, or this psychology fosters a scenario where God isn't required, and technocrats are idolised or even chosen because of their power.

Bregman in 2019 suggests acting powerless with powerful people makes people vulnerable, underestimate their own intelligence and less likely to strike back, then these people get treated stupid so they act stupid – catch 22, so the technocrats should take charge! Etc.

The power vicious circle

Stupidity is not an intellectual defect but a moral defect, power infects people with stupidity, or power requires the stupidity of another. People under power give up their autonomy and identity, so there is no limit on what the powerful can do.

You can recognise and resolve this by getting rid of the false image of being powerless, stupid etc, and understand that people are vulnerable to becoming victims of indoctrinated narratives.

Man without Jesus is evil, that is the definition of evil. Jesus is our infallible cornerstone. Jesus isn't about control he is about freedom, in pursuit of good, beauty, truth wisdom and practical help when we are stuck.

So, orientating ourselves around the basic tenets that "people are naturally good" without having a tangible a source of goodness other than oneself and one's own view, is no more than another variation of narcissism.

Pollution is a threat to sustainability microplastics etc, but Co2 is not a threat at current levels or on its current trajectory. The elite solution of Schwab etc is a dictatorship and imposition of a single government who rules over a dependent, constantly monitored, pharma spikes birth-controlled humanity.

This is anything but a dangerously naive dystopian Godless hell that will squander the ingenuity and innovative capacity of humanity.

It's up to us to find solutions to enable a successful coexistence, our success is dependent upon local and global cooperation. Socrates realised this in 400BC.

Social interaction is in our genes. Jesus died on the cross to have a relationship with us.

Xenophobia and conformity

Faith is easy to manipulate by inverting good as evil whilst taking advantage of our inclination for conformity and our altruistic behaviour. For example: the common objective narrative of "protection of others" combined the false narrative that we needed vaccines and needed to conform, make sacrifices too. But the road paved with good intentions turn to hell because the underlying drivers are false.

From the first false narrative the lockdown followed etc. The hollow narrative of protecting others is an excess of compassion and empathy without the foundational truth that Jesus provides. Or this claim is a corruption of the Christian faith. So, when this narrative is played to the desperate yearning/ remnants of Christian faith in the West, a corrupt version of love, a corruption of the first commandment. Or love thyself/ neighbour without loving God first.

People tend to have no empathy for people outside their circle, oxytocin primes us for herd behaviour within our circle, blocking out other factors or showing blind love and blind obedience enabling us to follow rules that are bad for us.

The divide between rich and poor can be explained in that way where we feel less love for others who are not in our group, or even divides us with the false labels of the rich are good and poor evil.

Those that we don't empathise with seem threatening, so empathy can divide as much as bring people closer. Empathy is compulsive and without thought and compassion is more rational. So, empathy creates a blind spot by disregarding the logic of others or being xenophobic.

Empathy without rational compassion is dangerous because it blinds and kills curiosity.

Without rational compassion or truth, we cannot see the bigger picture. So we should find solutions that are good for others as well as us to we can mitigate this pitfall.

So, we don't need a social scoring system that treats us as if we have no ethical compass: that just fuels the "power – idiot catch 22 loops".

Critical and lateral thinking

A more fundamental way of thinking outside the mainstream is required, and one that devoid of powerful influences and commercial interests.

The alternative psychology behind Suzanne Humphries view below is the body needs help in the same way as vitamin D to defend against virus's not harmful interventions into the body's natural immunity systems because the body fixes itself naturally.

"Suzanne Humphries from the International Medical School on Vaccination captures one side of this contentious issue with unwavering conviction. Humphries argues vehemently against the safety and efficacy of vaccines, stating that not only have ***vaccines never been safe****, and it is fundamentally* ***impossible for them to be so.*** *Her perspective is rooted in the belief that the process of vaccination inherently disrupts the natural functioning of the immune system. By introducing disease matter directly into the body, coupled with various chemicals and antibiotics, she contends that* ***vaccines place the immune system into a state of imbalance, leaving individuals more vulnerable to other diseases*** *than if they had not been vaccinated at all.*

Humphries' assertions challenge the prevailing medical consensus on vaccines, suggesting that the very mechanism intended to protect us actually undermines our immune system's natural defences. She points to the introduction of foreign substances into the body as a key flaw, emphasizing the unknown and potentially harmful effects of these substances when they interact within our systems. According to her, the idea ***of injecting disease matter along with other chemicals into muscle tissue is fundamentally flawed, creating a scenario where true safety is unattainable.***

Moreover, Humphries highlights a perceived dissonance between the purported benefits of vaccines and the reality she observes. She questions the narrative of vaccines as miraculous solutions, arguing that ***if they were truly as beneficial as claimed, the evidence would be undeniable and widely recognized. Instead, she suggests that adverse effects, such as severe illnesses and***

deaths, are overlooked or dismissed, with the public being seen as too ignorant to recognize these issues. *The historical context provided by Humphries regarding smallpox vaccinations further illustrates her point. She describes how children who were vaccinated against smallpox often became the sickest, suffering from severe ulcerations or even dying as a result. These historical instances, she argues, are indicative of a broader pattern where vaccines,* ***rather than providing protection, have led to significant harm.***

Humphries' perspective is not just a critique of vaccines themselves but also of the broader system that promotes and enforces their use. She views this system as prioritizing profit and control over genuine health and safety. The ***insistence on vaccinating infants****, despite the potential dangers, is, in her view, a manifestation of this system's* ***disregard for individual well-being in favour of broader agendas.***

Her stance is a call to question the safety and motives behind widespread vaccination practices, ***urging a re-evaluation of how we understand and approach disease prevention****."*

Source: Julian Elhaj

It is the development of system 2 thinking that enables us to think outside of the box like Suzanne. This requires a combination of empathy for those close and using rational compassion to welcome the challenge of others outside the group. In this way one can consider other views and the broader implications and play out the consequences and act and adjust actions accordingly.

Excessive empathy non thinking, or system 1 thinking, alongside ideology of any kind leads us to see ourselves as good and everyone else evil and threatening, even though they are just like us.

To restore this balance between compassion and empathy we need a fully functioning hippocampus to avert the technocratic ideological and elite centric empathetically fuelled tyranny.

The war on beliefs

We are amidst a global fight for our freedom. We face freedoms last stand, against our freedom of speech and our most basic beliefs.

This catastrophic violation of our heath should be the catalyst to open up pandoras box and resolve these matters, but there is a danger that the escalating ideology and narrow empathetic type non thinking that so successfully divides us could get more entrenched, just as the globalists would wish.

This is witnessed in the UN centrist globalists pushing to categorise and legislate that those that speak out against the falsehoods and tyranny of gender ideology as a crime against humanity. This is a fundamental attack on the basic tenets of Christianity too.

Melissa Fleming, head of the United Nations Department of Global Communications speaking at Boston University denounced all dissent, injuries, cancer risks, Mental Reset etc as conspiracy and baseless, gas lighting and fuelling the fear to maintain the technocrats lies based narrative.

This war is designed to indoctrinate us at the many intersections, climate, Israel, gender. But we can find the strength do the right things individually at times of crisis, especially when our children's future is at stake.

Very few people see this mess and our children are glued to their phones. They take in very little knowledge of what is unfolding, because for them in the online world "their rubber isn't hitting the road", because they are held in an unreal online scenario in part devoid from reality.

The solutions

It's a war against our ability to think, our vulnerabilities, blind spots and what unconsciously drives us.

It's vital to welcome new experiences, have open challenging discussions, invent solutions, show forgiveness, grace. The key to opening this door to be able to forgive is grace. Love and forgiveness are counterintuitive solutions. Yet as the bible teaches love is the only indestructible emotion and the foundation from which all else flows, so no other law is required.

Even Jesus could not heal without compassion.

We are also fighting for the freedom to talk openly and provocatively to cement relationships and develop the best thinkers and cooperation.

It's a fight for compassion – the elite technocrats don't care about people, yet compassion and love and grace are the solution to give us the higher meaning than ourselves that we need to discover a robust purpose.

That love should be rooted in Jesus, else our love falls prey to the myriads of other selfish destructive desires: power, money, self-etc. These diabolically corrupt good intentions.

There is an increase in empathy between people that see this threat. So, we must widen this circle with a view of more local economic control, to innovate and find local solutions.

Central solutions have never worked, because take "climate change" the solutions are completely different in the Middle East (keeping heat out of buildings and cooling) and the UK (keeping heat in buildings). A central myopic view co2 reduction is like Suzanne Humphries view of vaccines is the wrong solution. A growing economy in the developing world requires fossil fuels etc, so Co2 reduction the wrong metric. Co2 is plant food, in the same way that vitamin D is ignored as anti-viral protection.

A global government cannot work because the solutions are complex unique so cannot be centrally resolved. Innovation must be driven locally and collaboratively bottom up locally, whilst as Socrates noted, even in 400BC collaborating globally.

Social focused villages solve challenges locally and at the grass roots to be socially, economically, culturally sustainable through the productive interaction with inhabitants, to thereby find the best-rounded solutions, whilst preserving the best of humanity.

Eco villages, with medieval commons are an example of this but NOT 15-minute cities! In medieval times people tried to combine mutual success and the environment.

People given the opportunity are essentially cooperative beings and need the space and opportunity and love to innovate, alongside the state and market.

Social focused villages should be established without formal local structure maybe local centres pubs and community cafes where common objectives and collaboration can naturally form, and coalescence happen.

Heartworks café in Holmbury UK, is a fantastic example of this and how local social interaction can be opened to provide the platform to develop relationships, network, develop compassion and empathy within local village communities. From this we find people have the same basics requirements as we do and can very successfully think for themselves without AI.

Politicians, I think are likely have been worst affected by this indoctrination. This has influenced some absurd contradictions and actions that for the World Economic Forum conveniently leading to the victim centric populations of both persuasions trashing by blaming our democratic institutions and politicians.

This institutional demise, inadvertently fuels the justification and case for the technocrat's central solution, putting civilised society and democracy at risk of collapse. People need to understand how they have been indoctrinated and become anti-establishment too.

Multi-generational families could live together to provide support for older generations to create local family centric accountability. A Robust family is the bedrock of society. Family creates an excellent and safe place for older and younger members to preserve their identity, value and share experience. The current scenario is broadly rejecting the elderly as useless non contributors and a burden on society as well as the various woke ideologies attacking the family or corrupting its intended and real components.

Smart cities are not smart, they create stupidity.

The smart city was pioneered by IBM as a means of improving the way cities functioned, although they have been best deployed to enable authoritarian regimes, Putin, and Xi.

In China Ai controls biometric surveillance, 15-minute cities, social scoring, Digital IDs, CBDC, whilst they still operate under covid rules! So as Klaus Schwab suggests China is naturally the social blueprint for the West...!

Smart cities overlook the importance of people playing a part in developing their own local solutions. They suggest current cities lack intelligence, even though the reality is that cities have been incubators for incredible innovation ideas and intellect.

Optimising everything destroys everything and its every and the myriad of experience that makes a city wonderful. The efficiency creates monotony which kills our minds.

People are not attracted to efficiency and control; they want new experiences, good people relationships and community so that that they can feel a productive part of their society and realise and foster their purpose.

So smart cities rob people of their humanity and everything that is worthwhile to be human?

Community projects such as growing vegetables combine an activity with exercise as well as producing a valuable healthy source of food. This kills a few birds with one stone. There is a natural incentive to protect that local source of food, so it's part of the community's responsibility not an expedient, external purchased product from a supermarket.

Alternative views and where is the mass of opinion?

Covid 19 should make us more aware that the mainstream media are not to be trusted outright and the power of indoctrination. So, people should search out alternative sources of credible impartial information to help support their system 2 thinking and avoid indoctrination.

Paul Burgess robustly and independently provides this type of insight and lateral thinking on climate change. Covid 19 may have killed mainstream media, because media have become infected as well and corrupted by being funded by technocratic via investment from Gates and others. These parties have heavy investments in climate and pharma so are conflicted.

Social media and news tend to focus on the bad guys because that attracts interest. Sharing knowledge is a critical means of demonstrating the value of collaboration, and this is how the hippocampus is best protected.

22% of the German population have not been vaccinated so must have withstood some of the main indoctrination efforts. Yet many must have felt compelled to take the spike, despite knowing that the spike was wrong.

This is a high percentage of people that failed to fit in with the masses so we should be encouraged.

Most didn't receive their booster shot, so most must have realised that the spike was not about their health more about their indoctrination. The hope is that this every growing jury, will take advantage of this juncture address grievances.

The threat to resolving and undermining accountability to resolve this matter, is that the next engineered crisis, take bird flu is far more dangerous.

Bird flu that could cause 30%-50% deaths across the population. This is called disease x and is already being touted as a threat by the globalists.

As the facts and spike hangover continues, the globalists accepting a small aspect of responsibility for the jabs, to psychologically normalise their responsibility. Whilst fuelling the next phase of fear of the bird flu threat.

There is a vain effort to placate the masses by accepting some responsibility yet this responsibility is devoid of apology or accountability, and its failure just an opportunity for them to grab more power to mitigate failure next time around!

Until then we face an acute danger that totalitarian conditions could prevent us from getting a grip. This could leave the technocrats in power literally stepping over our bodies and interests.

Politics moving the masses.

Assuming those that see this issue can only convince one of 2 people each this could turn very quickly in our favour, but the Wef want their treaty signed in May in time for disease x!

Our UK politicians appear like Penny Mordant seem to be indoctrinated, because they are calling the blanket conspiracy theory technocratic narrative, so our democracies could be usurped, and the next crisis disease x launched. By not being held to account for the last one, that could get then off their accountability for the" Covid 19 crisis" too, by keeping the wheels of the vax Trojan Horse moving fast.

The question is if the parties that get the truth have been fatally tarnished, by being called Nazis or conspiracy theorists, the detractors of this form of socialism will naturally tend to be right wing. There is a danger that there may not party to vote for that has not been indoctrinated, because the alternative free thinking cynical parties will likely be identified and falsely labelled as Nazis etc.

Nothing is more dangerous to politics as public opinion, so maybe the new party should commit to empathy and rational compassion, because people right now feel ignored, unvaluable and threatened.

Finding the courage to get out of your comfort zone is essential for change. There is no other way to take the action required to escape the dystopia we are heading towards, so people must revolt.

The constant dripping of falsehood driven compliance, destroys the stone and people's integrity.

Evil tyrants rise by taking small, misguided steps that others follow and become complicit in, until it become too late. Germans in the 1930 realised this and the progress but became resigned to the fact that progress had gone too far for them to resist.

We must have difficult conversations and clearly state the danger.

What does good look like?

We don't want people to believe one thing or another, the outcome of the winningthinking® program is broadly members of cohorts achieving a clear blank mind from which to move forwards from, newfound skills, reconnection with others and great relationships and a feeling of dependence upon others. This is the basis of the new strength required to face and address this challenge.

From this healthy place and feeling of being valued, people can form their own views and regain themselves.

A focus on believing the accepted wisdom, is exactly what the technocrats and ideologues want. And pushing an alternative narrative may not be much better.

People will survive only if they have a profound sense of their identity, individuality, and the protection of the right to develop their own strong personalities, in the context of co-operation and compassion, devoid of self-reliance and narcissi. Social power and relationships are key to defeat technocratic oppression. This is why they hate the church because Jesus died for us to have a relationship with us, relationships matter that much.

This reverts to Suzannes view that we don't need vaccinations and climate ideology, to protect ourselves from ourselves, we can rely on natural immunity and natural intelligence and collaboration to develop strategies to protect and reconnect ourselves with the environment. This is counter to 15-minute cities that will destroy our will and create a meaningful connected loving graceful life free of ideology.

We must coalesce and band together, although as I describe before and after this chapter technology fuelled weaponised narratives planted in volume and fuelled with AI is fighting against democracy.

People are uniquely capable of self-reflection, learning, compassion, grace, and love – this is what good looks like. However, technology and humanity are driven by completely different paradigms, inputs and outputs so are like mixing oil and water. AI is completely devoid of ethics.

The worst narrative is that “the human is a fatal aberration of and threat to nature”.

Time is of the essence.

It will not be cruelty but the threat of the weakness, docileness and lack of accountability and lack of responsibility of man.

So, we must find our inner strength, that preserves our individuality, opportunities for new experience, and avoiding the mundane soul-destroying reality of 15-minute cities.

We must value each other for who we are and welcome different views and develop forgiveness, grace, gratefulness, and love as our modus operand.

These are Christian virtues and appreciation is a gift from Jesus, the alternative is the curse of “more” which equates to valuing nothing and like the technocrats being satisfied by nothing so pushing on regardless.

People with Jesus are good. We probably need to influence of 2 others of the threat that we know, because most people now realise that the indoctrination of the jab was a mistake and that they were duped by this technocratic pharma aspect of the Trojan Horse that has no underlying interests in their wellbeing.

Politicians caught in the middle of this indoctrinated mess will have to eat humble pie and take a lead to resolve the matters they have funded and rolled out.

This leaves a skills gap to forge this new capability, because left to their own devices the lazy victim centric habits of living out this ideology will take a long time to unwind. People's ability to have difficult conversations is impaired. People must face the facts that one's beliefs could have been lies; this will cause people to get angry and defensive and put the mantras of taking offense and hate speech to one side.

Now is the time for this type of winningthinking® capability and awareness to be led top down in organisations and actions being bottom up to suit the needs of local communities. People must speak out and revolt if necessary to avoid becoming complicit, because being complicit fuels the diabolical leaving the good people to become part of the problem and fuel this evil too.

This change would cause the technocrats to lose their power over us, make our lives worthwhile and pleasant and restore the beauty of humanity.

Jesus is all about beauty, truth and good as well-being as the source of love so let's give up our will along the way to Him. The alternative is various iterations of narcissi which unwind into a different type of intersectional division and reliance on self, which manifest in the "Western post Christian emptiness" that now dominates many people's existence.

This logos is fuelled through intimate connections between team members and society, thereby realising and rediscovering what matters. It's by taking different perspectives and realigning actions that this healthy winning thinking thrives.

The tough thing to realise is that when things get tough to focus on helping others because a focus on self-reliance, causes and accelerates a systemic collapse. This and other robust outcomes are experienced by delegates in chapter 1 of the winningthinking® masterclass.

A desperate attempt to cloud seed to block out the sun May 2024 UK SE England

If we let the technocrats use these weaponised narratives to impose cloud seeding, uncontrolled immigration and mind destroying virus, upon us the next logical step would be to raise the bar and launch disease x and develop the virus to become transmissible to humans.

These questions are being asked now because disease x has reached Australia.

> "Mr. Leake and Mr. Hulscher alleged that several months after the start of "this gain-of-function attempt" in 2021, a new H5N1 virus (clade 2.3.4.4b) appeared in wild birds in the United States, leading to the current outbreaks."
>
> **Robin Westenra**
>
> Gain-of-Function H5N1 virus reaches Australia
>
> SEEMOREROCKS.SUBSTACK.COM

Tom Pickering
Woke Culture And
The Mechanics Of Tyranny
The Evil of
Silence
The Evil of Silence
Woke Culture and The Mechanics of Tyranny
The "woke" mechanism is the same one used to escalate the Nazi party. Very few people saw what was happening. Even less raised concerns.
The Evil of Silence is authored to inform the debate and provoke consideration and recognition of the consequences of the complex subject that is woke.
Nothing drives authoritarian control and victimhood more than Postmodernism or Woke Culture, Slavery, and even Nazism. Even social media has become weaponized. Such global platforms allow dialogue to be presented and then unwittingly manipulated, leaving large corporations and even the UK Government in precarious positions. Pander to the woke culture or face reputation-damaging criticism?
When dialogue stops, physical or psychological violence starts.
The alternative? This book shares mechanisms to raise the awareness of and sharpen our skills to resolve tough issues in an increasingly disparaged environment. Otherwise, we are leaving ourselves and the very underpinning of society as we know it wide open to authoritarian control. The psychology that causes a company crisis is the same destructive psychology that fuels Woke Culture and Nazism.
The Evil of Silence precedes the debate presented in Tom Pickering's latest book, Why Can't We Stop! How to Avoid Menticide: The 2020's Cause of Acting & Thinking Recklessly. Written by an unapologetic, quick-witted, hands-on business turnaround expert, portfolio chair, CEO, former Tech Silicon Valley Exec & MBA Masterclass Leader, the book offers a pragmatic take on the consequences of staying silent in a world that is in desperate need of social change.
Now is the time to become aware of the misleading guises of woke culture and speak up.
Barcode - Do not remove

The Evil of Silence: Introduction summary.

The Evil of Silence: The Common Denominators of Woke Culture, Nazism, and Slavery

The “woke” mechanism is the same one used to escalate the Nazi party. Very few people saw what was happening.

Even less raised concerns.

The Evil of Silence is authored to inform the debate and provoke consideration and recognition of the consequences of the complex subject that is woke.

Nothing drives authoritarian control and victimhood more than Postmodernism or Woke Culture, Slavery, and even Nazism.

Even social media has become weaponized. Such global platforms allow dialogue to be presented and then unwittingly manipulated, leaving large corporations and even the UK Government in precarious positions. Pander to the woke culture or face reputation-damaging criticism?

When dialogue stops, physical or psychological violence starts.

The alternative? This book shares mechanisms to raise the awareness of and sharpen our skills to resolve tough issues in an increasingly disparaged environment. Otherwise, we are leaving ourselves and the very underpinning of society as we know it wide open to authoritarian control. The psychology that causes a company crisis is the same destructive psychology that fuels Woke Culture and Nazism.

The Evil of Silence precedes the debate presented in Tom Pickering’s latest book, *Why Can’t We Stop! How to Avoid*

*Menticide: The 2020’s Cause of Acting & Thinking Recklessly.c*Written by an unapologetic, quick-witted, hands-on business turnaround expert, portfolio chair, CEO, former Tech Silicon Valley Exec & MBA Masterclass Leader, the book offers a pragmatic take on the consequences of staying silent in a world that is in desperate need of social change.

Now is the time to become aware of the misleading guises of woke culture and speak up.

Let us get our heads in the right place.

You may think that Woke culture is just about anti-discrimination, but Woke culture is rooted in authoritarian control, as it is rooted in victimhood.

The same methods deployed by the spread of woke culture fuelled the rise of slavery, Nazism, China's control of Uyghurs, and Big Tech control of the dialogue, perfectly facilitated and weaponized by social media. Even China's policy of moral disarmament is strongly comparable to woke culture, in terms of methodology. Let's explore: *China are the experts on silence* Sun Tzu's *Art of War* observes that perception of power has a power of its own and that the most elegant defeat is the one incurred by the moral disarmament of the enemy...

In short, Xi Jinping's command group must be aware that many windows are closing for it and that time is not on their side unless they can persuade the Free World to self-harm sufficiently that we disarm morally. For this, there is evidence of intent and, unfortunately, of current success...

Ghost attack is described thus in two stages: Stage One: "Commit non-attributable, or feasibly deniable, acts of war that advance your own national power and physically harm your adversaries, but do not fit traditional legal standards to mobilize them to a formal response." Stage Two: "Then foment chaos, advance the narrative of your innocence, and underline your adversary's incompetence through information warfare."

What is the link between China, Slavery, Woke culture, and Nazi Germany?

Very few people saw what was coming with the rise of Nazi Germany and those that could be in the minority, could see it but everyone around them was tragically wrong. These people have already seen enough to reject it. It was NOT the masses that took down the resulting tyranny in the Gulags, it was a few extraordinarily strong people who would not conform to the system.

Victimhood and method.

Woke culture derives its power and moral high ground from those that adopt the victim persona and the defence of that as the only opinion that matters, providing the ideal platform to attack and broaden the woke agenda. Yet psychologically, victimhood is the unhealthiest perspective to take because it conveniently drives blame and externalises responsibility and stasis. Defending these victims is central to woke strategy as well as using the platform to pursue their broader agenda to undermine society. Victimhood was the tool used by the Nazis blaming the Jews or Woke culture routing out ill-gotten gains or racism, when in fact they are creating a cast of woke vs non woke which is far more sinister and pervasive.

The woke power is derived by attacking those as being amoral for not supporting these victims levelling subjective accusations and mobbing on social media. This leaves those that have a moral backbone and courage in the firing line for challenging the patently absurd associate strategies such as banning the word “mother” and replacing it with the word “birthing partner.”

Marginalising people in this way leaves them without support from their friends – driving real fear.

The dialogue is transmitted on social media and then unwittingly manipulated, leaving large corporations and even the UK Government in fear of being criticised if they do not pander to the woke culture of taking the knee in solidarity of Black Lives Matter and following the control of speech. Surely, this is a sinister act of total submission and overruling, which distracts from the very reality that ALL lives matter.

The thing about controlling speech is that humans must align their beliefs to cope, so controlling speech manipulates beliefs. This is known as Cognitive Dissonance and is not far off the concept of the dystopian matrix where people just become part of the program fuelling the ever-increasing tyranny.

This is part of a more sinister plan that has a common method across Nazism, neo-Marxism, UK unconscious bias training, China's psychology, and identity destruction of the Uyghur People.

Why should we be concerned, or will it fade out?

Does moral panic burn itself out, and recede by itself? It does not seem to recede on its own, it is touching everything Big Tech etc, anti-group prejudice. Many suggest woke culture is just about equality, the "Systemic Racism" label for example does not stand on its own, it is part of a system of similar ideas...

Critical race theory.

Language is misused by compelling certain dialogue as part of the ideology.

Identity too is complex and negotiated with others, not a simplistic group conveniently exercising power over others.

The label of "Systemic Racism" is impossible to defend yourself and is a devastating accusation.

Am I right opposing this? Systemic means pervasive everywhere, the central tendency of the system. This is a negative generalization and simply wrong when the broad central tendency is in fact cooperation to a good end. Racism is so vicious, propagandistic many cannot put together a reasonable counterargument, so it sticks regardless of if it is justified.

There are elements of racism, the legal system and we should acknowledge it specifically. But the argument is intended to tear down the west. People are ethnic centric – leary of different races, tend towards their own isolated groups, and there has been a real risk created by the spread of infectious agents through global trade. Slavery is a universal human tragedy regardless of race, although slavery was abolished first in the UK, then the USA.

The dialogue about wealth and privilege has been "Ill-gotten stuff so torn down", and society is corrupt right to the core. This has a common link with the anarchist anti-establishment aspects of bitcoin, and technology as is described later.

There is a linkage between ideas – there is an entire ideology, based upon inequality of outcome used to selectively justify unfairness systemic discrimination, and bigotry.

Organizations that support BLM support abolishing the nuclear family, defunding the police, and believe capitalism is evil. These institutions are far from perfect but have taken hundreds of years to build and are far easier to wreck than rebuild.

Language is hijacked, dialogue bipolar so geared to conflict, so you are either collectively guilty or innocent. Good people who stand up to the untruths are made examples of as those who do not fit. This sends a message to people who do not fit - this will happen to you – fired and silences those around the accused too.

Who do you have to appeal to, who is accountable?

Social media and tech have enabled information to be democratised, no longer hard to get information so the tendency is to take sides and "ra ra" for your team.

If you step out, you cannot rely on people supporting you – no protection. There are some examples of this dialogue below and how to make sense of it and react.

Education is tending towards propaganda and untruths rather than a healthy and open exploration of ideas.

It is difficult to say what's driving this, the power is driving ideology, which conflicts with the reversed caste system for the marginalised.

A virtuous position taken is standing up for the marginalised or victims in society, although accountability is hidden by the crowd, laughing about the bloody revolution, even if it is warped and twisted.

People submitting to Woke culture are not bad. They are only doing the best they can, and open dialogue is key. Beware, history is clear on this: when dialogue stops the only alternative option is physical or psychological violence.

Woke culture shows the common concern for the disposed of, which goes back to the well-intentioned black slang "are you woke to this?"

This "moral high ground" is being used to tear systems down when the current scenario is the best antidote to these common problems.

Why is this done with so much force we consider we are so remote from the violent revolution? It is a luxury to flirt with these ideas, yet the progress humanity has made is a miracle to live in and Woke culture shows no gratitude or acknowledgement of beauty.

Utopian thinking is first to break it to pieces so that something interesting might happen – so that they can have their way. We maybe offer too much decadence without any adventure, so the Woke advocates probably desire for recreating meaning, purpose, and adventure.

It is mainly faith that makes people stand up against this. What are people willing to sacrifice - their jobs, careers etc? Most strength is faith anchored upon something deeper. So, faith groups are sure under attack and have no credence in Woke culture, so robbing people of and discrediting religious beliefs and their deep connection is key. Woke culture is a religion.

Where is the moral frontier, where is the group interested in telling the truth to take on these people searching for that romantic adventure?

Victimhood, someone's gender or identity does not define whether something is true or holds more sway than another view?

Are we overstating or underestimating the danger?

This could be the freest society ever known, but if people do not speak out will we be living in a totalitarian society? Increasingly people are saying "I could never say this out loud" When this is widespread it becomes a dawning of a totalitarian state, defined by a split between public vs private viewpoints. For sure, people did not say what they needed to say to prevent the Nazis murdering the Jews. Say what you think carefully because the alternative is far worse.

"Every great man is the actor of his own ideal" Nietzsche. Acting out our ideal is how we find our strength and purpose

Woke culture prevents many living by their simplest rights or ideals. We cannot say common sense views out loud: such as "men and women", "mothers," "the USA is a good place", or speak against an ideology hinged on and defined by exploitative power and institutional corruption. This is wrong.

An intolerance for free speech is enough of an indicator – rather than being wrong now.

Where are the broad thinking people who care? They have no space to operate and are entitled to a different view?

The ideas lead to literal violence and more blame such as the "national guard caused the rioting and peaceful protest."

Should corporations be moral actors? Rather they are becoming "offence archaeologists", this for sure creates a hell and must be a bad use of time – versus producing a product?

Moral courage.

Do not apologise if you have not done anything wrong. The bold people whose roles are centred around doing the right things based upon searching out the truth, they are getting fired, so have nothing to lose. Woke culture is making their jobs become impossible. The danger is that people become a half version of themselves, very damaging for those that have a clear moral compass and care.

This trashes integrity and creates a husk of people. The optimistic align actions with beliefs and find their purpose.

Institutions are bending to meet this skewed Woke morality to achieve heroin hits with the clients and or readers and to avoid being attacked publicly themselves. The question is this: are organisations tailoring to the audience and creating genuine conversation and sense of perspective. That said, it is important to have healthy counsel, otherwise you can shape everything and become a parody, so you need to build a healthy counter ecosystem around you, not ECHO chamber!

What is dehumanising or surely the most offensive form of attack is being victimised or being personally attacked for your views. Whilst the young have an apprenticeship in their early careers, after that they learn how to speak, but if they take these narrow ideas from their education, this makes it impossible.

What good looks like and where is the moral courage from our younger voices?

We must take a view of what is good and wrong to create high standards, otherwise, where do these high standards come from? This blurring of right and wrong and lack of leadership courage is diminishing or destroying the high standards built up over centuries. We all need a safe space to enable us to cope with this harsh Woke criticism or be clear that the critic has a genuine wanting for the best. Trust your heart.

This open mind requires us to put our egos aside and listen to them. The bright big thinkers will keep what is great, and study what works by mapping it out.

The discipline to achieve the basics of your craft is hard enough, let alone when the moral standards of your learning environment are questionable. There must be an allowance to deviate from common thinking where possible and take new action.

Integrity is defined by those that can swim upstream or have an element of disagreeableness in their character. These are the people who tell you what they think or an annoying superpower.

We must separate the Identity labels vs ideas else we will lose institutions capable of training people. Letting go is hard because Woke culture is a comfortable place. It has required great courage to build great organizations, it is tragically easy to pull them down using Woke culture because you are immobilizing the good open thinkers with a strong moral compass to be replaced by an anti-establishment, anti-nuclear family tyranny that doesn't seem to have a constructive plan.

So, the intention of this paper is to offer an apolitical view and apolitical solution, although you will find yourself labelling the views expressed that are part of the journey and shows the flaws in our psychology.

Some aspects of psychology are clear though which I will try and articulate, again many will disagree and that is OK.

What is the link with Company Turnaround?

Why is this specifically of interest to me? I have studied the psychology of company turnaround, using MSc level psychology methods. We know how and why business psychology corrupts strategy. Woke culture deploys many of the same destructive tools, so is remarkably familiar.

Furthermore, we have much experience of creating the antidote, captured in winningthinking® game-changer program which we have defined, deployed, and crafted over thousands of hours in 200 critical business scenarios, over the last 15 years.

The proof of the method is tested and proven to overcome tough intransigent scenarios where people have no idea of the issue – blindsided and immediate new action and immediate cash and people-related results are required.

Let us open the dialogue, flush out the issues, take a new perspective and jointly take new action to restart a constructive conversation?

This book boldly opens the lid on the issues and understanding of the mechanics of woke culture, the skills needed to break any polarisation or conflict and raise the awareness of the skills to engage in productive dialogue and productive relationships.

Consider these perspectives:

"When talking stops that's when wars start."

"A victim has lost the battle of their life."

"Victims handicap themselves for life."

Establishing common ground.

Anyone would agree that in the context of letting us say “diversity and inclusion,” it is important to value different views and bring them into the light to keep society healthy and avoid the perils of group thinking and tyranny and silence?

The below explains the corruption of woke culture, which started as perfectly well-intentioned black American slang, “are you woke to this?” or are you aware of this? The main motivation in sharing this paper is the damage done due to the current fear of speaking up and silence.

Creating common ground: As a line in the sand and frame of reference – what might good look like?

“Acknowledging people for who they are and taking an interest and value in a different view.”

What could stop us from achieving this?

We need to provide people with new opportunities.

And act out of a love for others – by putting them first.

Why is woke culture a concern and where and how it is corrupting broader society, why it will get worse unless it is addressed? Right now, woke organisations like Stonewall, BLM and others corrupting businesses with their mantras and metrics, based on half-truths or anti-truths and through a new tyre of dialogue are successfully silencing their critics.

I think it is fair to say Woke culture is creating a dystopia and playing God. This book describes the mechanics and deception of Woke culture so that you can recognise it and develop the skills to keep in conversation with.

The background is its ideology of suppression of people by hierarchy and tyranny. A tragedy is that 1. IQ and 2. conscientiousness is the driver of success, and IQ is very unfairly distributed.

This seems to lead to a woke focus on trashing merit when in fact merit and or good character is the fairest means of promoting people in organisations. There too is little evidence of unfairness, in recruitment or selection, in fact, organisations for the last 30 years have bent over backwards to become fairer to marginalised groups. This is the basis of the ideology behind systemic racism. It is an assault on merit in society, and in fact, beauty, whether culture is attacking marriage, motherhood, heterosexual sexuality, and searching out anything negative to fit with the self-righteous woke mantras. Family, motherhood, male, and female identity are all key elements that are the backbone of society, and the basis of what people care about and fight wars to protect.

On the contrary, though, competence is a good way to stop prejudice and alternatives are far more unfair.

Furthermore, anyone brushed with speaking out scandal is at risk of becoming marginalized and isolated by even their peers. The woke accusations tend to be ambiguous and labels such as racist sexist, offensive, ill-defined rather than the facts are distorted or deleted to fit the woke mantras.

The current environment.

In tough and uncertain times people will search out and follow mantras or generalizations even if there are flawed. They can create a sense of belonging for the marginalized. The level of shock and loss of meaning following the pandemic is not dissimilar to the period after WW1. This makes people vulnerable to charisma and bold propaganda and statements as a route to purpose and certainty.

It may be the case, therefore, that those with lower IQs, or victims, find these woke philosophies very appealing and provide an opportunity with:

1. the simple attack language

2. a "virtuous" base from which to defend, and

3. gaslighting tools to undermine the credibility of the counterargument to wreak their revenge.

This type of strategy avoids the need to have dialogue, face untruths underlying their thinking or supporting virtuous positions, whether directly or the third parties perceived unfair treatment as "perceived victims of oppression".

Victims are defined as those who blame. There is nothing stopping anyone from acting?

The paradox is whilst Woke culture claims to stand up against tyranny, its methods are creating a far greater tyranny.

Woke attacks can happen to anyone, so we all should be fearful, and they are a direct attack on reputation, severe disproportionate punishment, as well as Orwellian unconscious bias training.

The danger is that no one will ever say anything controversial, so ideas or truth will never be discovered. Unchallenged thoughts become facts – this bends good people out of shape.

Woke culture attacks the family, sex, Christian culture, etc. These are the fabric of society; it is oddly hard to defend because we take these pillars of society for granted.

Flawed labels and mantra-like male patriarchy, systemic racism, and equality cause this corruption and silence. You need to be able to identify these and how they drive actions and behaviour.

These labels create blind spots, wilful blindness, which prevent reaching the truth and encourage those with low integrity.

One of the most basic flaws with the woke equality mantra is that it is outcome focussed, for example, many more men are in prison than women, so should we convict more women? Equality of opportunity should be the focus based upon sensible criteria. You can see the racist or "positive discrimination" that is taking place. Positive discrimination has negative connotations for others. This is a good example of how flawed mantras create a never-ending cycle of corruption and unfairness deviating further from the truth and screwing up the good people and thinkers that value the truth in the process.

Social media facilitates bullying and enjoys silence because it protects their platforms. People naturally fundamentally cannot cope with the variety and volume of different views that social media creates, so silencing controversy and woke culture is convenient to support their growth and complicity in untruths or narrowing the debate. A specific example of this is Facebook silencing those suggesting the C19 emanated from the Wuhan lab.

You will notice the dialogue refers to and attacks other groups and attempts to silence not value other views, which creates conflict. When people cannot speak up ultimately their only other option left is violence.

So paradoxically, the woke culture is a philosophy driving culture and tyrannical power, forcing people to act against their own interests. This is the antithesis of using your will. This is very unhealthy, causing extreme stress in people that really care and they will either collapse or leave.

The power struggle between groups leaving no individuals with no power, it's an unconscious confession of cynical Machiavellian, narcissistic, thinkers and thoughts and because the expert on woke culture speaks on behalf of a third party or their woke group they can bully without repercussion.

This group ideology power motivation is attacking western culture, and human endeavour, and takes fundamentals all the way to the bottom. It ends up with war or loss of life, for example, Marxist methods led to the deaths of 100m people last century. Woke culture is very closely related to a neo-Marxist philosophy, neo because it’s a theory.

A theory too is difficult to argue with because it is not factual.

Conscientiousness and intelligence are drivers of success, the importance of intelligence is too tough to digest. Improving hiring processes and hiring the best people often punishes managers because someone else takes the benefits. MBTI is an easier tool to use because it avoids many of these conflicts.

Let us think about woke culture and its corruption EQ is not contrary to IQ or reading other people is the most complicated thing to do.

The toughest aspect is that intelligence is not distributed fairly, but the flaw is everyone can be trained to do anything, this is false. "Woke defined personality defects' hobble the best people too and will not be overtly considered.

The most capable are capable to take the action to most benefit the less fortunate, yet woke culture is a great tool to attack the most capable people who have strong values and capability to create the environment to best help the less fortunate.

I noticed in my first senior executive position, staying on the path of truth or honesty requires real capability, yet it is far easier to obfuscate the truth.

Maintaining higher moral claims is dodgy ground because we are all corrupt, so leaders must not shut down the means to improve and search out the truth discovered through free speech. We as leaders must encourage the contrarians and use the contradictions to challenge our own thinking.

What are the key skills to maintain healthy dialogue?

There is an underlying challenge which is to become aware of the skills to identify flawed psychology, remove blind spots and give the tools on how to depersonalise the discussion to resolve the right issue. This is the crux of the winningthinking® process which will be shared later. True wisdom is to recognise we are the problem and to develop the skills to see and do something about this.

Our well-being and checks and balances are also highly reliant on others.

People are very capable of acting out untruths: these will be the things that you get angry about when challenged. The focus is not about the ego or being right/ wrong, it is about developing the skills to have the conversation.

Ethics do matter. A good test is whether you would bring your sexual behaviour, say multi-partners or use of pornography. into the light in public, acting out of the lie. Pornography is not good, sex within a family stabilises our families' societies and psyches to become our highest way of being.

There is a challenge when technology focuses on "expedience vs meaning" as well technology fuelling addiction e.g., pornography. Malevolence is a universal problem for all of us, motivated by power and ungratefulness and in a tech environment - addiction.

Knowingly saying things that are false keeps you in fog. Truth serves the truthful and good, falsehood can be more expedient and so more tempting. It requires faith to conquer this.

True wisdom is to realise "we" are the problem, not "them".

There is no fun or humour in woke culture, and fun is a real leveller of assumptions and facilitator of discussions, yet 86% of 500 voters in a recent poll said they valued the emojis and funny stuff at work.

If you wish to challenge some of these thoughts or assumptions, you are welcome to research these yourself because I cannot think of anything worse than accepting my view blindly anyway and what is the value in that if you have not thought these challenges through yourself?

Woke the definition.

Verb: alert to injustice in society, especially racism. "We need to stay angry, and stay woke"

“Stay woke became a watchword in parts of the black community for those who were self-aware, questioning the dominant paradigm, and striving for something better. But stay woke and woke became part of a wider discussion in 2014,

immediately following the shooting of Michael Brown in Ferguson, Missouri. The word woke became entwined with the Black Lives Matter movement; instead of just being a word that signalled awareness of injustice or racial tension, it became a word of action. Activists were “woken” and called on others to stay “woke”.” Source Merriam Webster

Vague accusations, wreck identity, and good people torture themselves through these accusations.

The mechanism of how we receive and process information.

The environment

We do not operate well in the context of increasing complexity, stress, vast volumes of information, all fuelling uncertainty.

So, the response, time, and mental space to take on others' views is limited, and the appetite to take on different views far less so. It is far more likely that we might get stuck in the fight or flight mode where new information appears as a threat rather than an opportunity.

The tragedy is that our brains are hard-wired to search out negative things as a survival mechanism so do extraordinarily little with the positive. Negative also triggers dopamine and serotonin – these feel-good hormones that drugs like alcohol trigger and which lead to addiction.

You will notice that woke culture encourages and searches out the negative to undermine reputation, an attack without any implicit moral standing or right to moral high ground. The paradox is that the woke culture bases its fights on perceived victimhood and inflames the issues and shuts the dialogue using negative generalizations and distorting and deleting facts to fit to fulfil these flawed mantras.

This strategy is attractive because it creates certainty (although flawed) and is a means of controlling a volume of information without having to think about it, take responsibility, or do anything different.

The attack tool is the generalizations that cause the facts to be distorted and deleted to fit a woke racist, sexist personal mantra.

Taking offence.

Taking offence on behalf of others is a distortion because no one is qualified to take offence on behalf of others, they must speak for themselves. Taking offence on behalf of others is counter to the woke culture of "you don't know."

Taking offence by adding meaning or by implication, for example, woke culture often labels a challenge to the male and female sexes as transphobia. This is not true and does not follow. In fact, feminists would practically argue that gender undermines the sexism legislation.

Personal matters because it is an aggressive personal attack not an assertive discussion of the issue – that is the definition of aggressive vs assertive approach to deal with the issue. This leaves the check of the validity of the woke issue untouched and unchallenged.

At its most basic, this very primitive lazy thinking is incredibly attractive for the marginalized and a powerful rallying tool for them to revolt against the masses that they perceive are oppressing them.

This is a victim mentality rather than taking ownership, and the reinforcement from others is self-fulfilling. Clearly, in this process, there is no thinking or learning and it does not attract the best minds. It is great though to fire up those that might feel shame or guilt for their inactivity, create purpose and mobilize them into angry action.

The second model option below requires courage to speak out against because you will be hit by the masses for breaking their rigid woke rules. Rules that it is fair to say form the basis of the woke tyranny.

To make sense of the woke dialogue, try to identify and call out generalizations to understand their impact and outcomes: the distortions and deletions. This will be the basis of and distortions and deletions and assumptions. These prevent focus and getting to the specifics to reach productive dialogue because the general self-righteous personal labels of racist, anti-inclusive, etc are the basis of conflict and so reinforcing blind spots but weirdly enough are wonderfully comfortable.

Because the woke mantras are flawed, the response will be seen as threatening and receive an angry response too.

The tools deployed to defend the indefensible are gaslighting, distorting marginalizing, labelling, such as "you will never know." These strategies have become pervasive: Stonewall metrics have been embedded in 850 organizations including GSK and others and ultimately lead to the whitewashing using the tyrannical tragedy of unconscious bias training that employees are forced to undergo. This is akin to the tragic cruelty and re-education of Uyghurs in China.

Those that do not care, do not value the truth and will go along with it, those with strong moral fibre and views will become rightfully angry.

Therefore, the woke dialogue fuels and creates distortions, deletions and enables lies, and prevents getting to the truth. This is how ethics can deteriorate without any reference except for the pervasive distorted justifications in the name of "selective personal freedom" for example the EU's move to propose abortions at any time during a pregnancy is an individual's right and anything else is seen as fitting the woke "oppressive" mantra.

The relationships in the woke model are flawed too as the individuals and the individuals following the mantras are not learning from one another.

The tyranny and reduction in free speech is getting more intense.

Processes that create increasing amounts of rework indicate where things are not working.

BLM is a classic generalization that has led to protests: conflict and escalating violence to ensue by fitting facts to the BLM mantras.

The murder of George Floyd was carried out by Escalation fuelled by the generalizations, for example, BLM non-white police, and whilst abhorrent there was no conviction for racism only murder.

Mental health is deteriorating; the tyranny is distorting an individual's actions and beliefs – causing people to lose themselves and become stressed.

Impossible quandaries are being created: Gender redefined athletes in sport creating unfairness; new and increasing definitions of gender trumping sex creating confusion in children and contagion in girls; People are forced to think that homosexuality any form of sexuality, etc is OK and denied their view; Heterosexuality (the crux of society) is being marginalized so "H" and does not appear in the LGBT+ lists.

Male and female strengths and roles as men and women are the backbones of humanity.

The outcomes are segregated outcomes and based on race – such as black-only investment groups, or mandating boards to employ minority groups on boards, this is a distortion and racist?

Must have black representation "we are the only ones that understand equality" – is a policy based on race and is racially biased against non-blacks?

Being offended trumps everything although what is the definition of offensive? Is offence just a distorted perspective of what a challenging alternative idea to your thinking entails and feels like? It is important to take on board a challenge rather than rejecting and taking offence. If you watch the woke dialogue offence is the lazy woke option to fit with the generalizations and stall and maintain the moral high ground.

A more robust strategy must be to make the choice to "THINK" or how will woke culture shift or develop? Yet there are those who are being personally attacked by the lazy corruption of woke culture. Leaders must lead and use their ingenuity and ability to think through complex issues and relentlessly search out the truth.

People who find the woke strategy attractive may not have been given opportunities, so the victim badge is convenient and attractive means to divert attention from them taking accountability or responsibility for themselves or their lives. Many have just given up on being proactive so fight back without taking ownership. Their woke victim-centric strategies conveniently obfuscate the truth and ownership, online echo chambers and mantras fuel the tyranny.

This is a comfortable strategy because it justifies their status quo, externalises the issue and justifies doing nothing.

The key is to provide opportunities open to all, but those opportunities must be fair, open to all or a selection based on race, sex, etc.

"I look to a day when people will not be judged by the color of their skin, but by the content of their character." Martin Luther King

The paradox - unlawful censorship.

"People often forget that the very concept of political correctness or woke culture was invented by Stalin's Soviet Union. Now... that social media companies have decided to become "Glavlit; "that is to publish only material that is supposedly truthful and passes its "community standards" This is more like the former Soviet Union than the United States under the First Amendment.

So social media companies should stop censoring other speech based on criteria of supposed truthfulness, "community standards" and other such questionable criteria that are subject to political, ideological, and other biases.

Who wants censorship other than for material that is already prohibited by law?

But if the social media companies persist in censoring, they must apply a single standard to everything? The current social media platforms have the worst of both worlds: they censor material that is neither dangerous nor necessarily false, and then permit material which is both highly dangerous and demonstrably false.

Do not underestimate the outcome of Marxism and where it leads. Last century, over 100M people were killed under Mao, Lenin, Marx, and Trotsky and 60m people by the far-right.

The neo-Marxist common denominator was group thinking trumping individual views and an ideology based upon the perception of oppression and victimhood of the working classes: this is called neo-Marxism.

Violence happens because neo-Marxist ideology cannot cope with the diversity of thought and the increasing degree of control required to oppress regimes.

This oppression kills enterprise and creates escalating poverty. Not all nations start this way, North Korea, with the Kim regime, created such tyranny with many dying of starvation in a few decades. It is interesting how in tough times people get duped and become vulnerable to such flawed ideology. This is incredible, so it is important to explain how this tyranny could take root in the UK. It is important to make sense of, how to recognize and reverse this woke blight to save our free speech and democracy.

Woke culture started well. It was simply Black slang for "are you woke to this" or ``aware of this." Endlessly self-righteously searching out negatives is a road to hell and the good intentions have become corrupted. It does also fuel an overdose of self-virtue and moral standing when combined with the woke attack methods gracelessly levied on its critics too.

The characteristics of woke dialogue.

The axis of woke culture now focuses on: Undermining the basics of society: families; motherhood; male and female roles; undermining beauty and humour.

You will find the dialogue centres on personal, sensitive, emotive issues such as race, sexuality, minorities, oppression, taking offence, self-righteousness, and victimhood.

This clashes with other groups – faith, leading to relentlessly searching out negatives from history to discredit the opposition and win the moral high ground.

The moral high ground is based upon "redefined truths" that manifest themselves in generalisations and flawed labels such as "Systemic racism," "Sexism" or "Trans hater."

These are vicious accusations that cover up the underlying specifics for fear of being accused of being racist etc.

The racist or offensive are ill-defined terms, so once accused, the accusation is indefensible, the outcomes are polarisation of views with little debate increasing anger and reducing grace.

Social media support/mob bullying control – fear of speaking out.

People losing jobs for diverse religious and other views.

People speaking on behalf of others – like the F1 girls who lost the jobs that they loved. They were thought by the feminists to be incapable of making their own judgement. We should celebrate beauty without denigrating it.

People labelled as scandalous are therefore an overt risk to employ.

The self-righteousness of the ineffective masses silences the genuine heroes who dare to step out and say what they feel is right, so the wrong people are thriving in the uncontrolled silence.

Direct implications / emerging concerns & hypotheses

An increase in male suicide - these are the active men and the heroes who benefit everyone.

LGBT labels create confused kids leading to contagion in young female suicides and silencing of those who try and help them make choices for themselves.

The victim and the person taking offence have the upper hand in the dialogue regardless of merit.

Taking offence is the perfect means to avoid discussing difficult critical subjects and the specifics, not the label.

Cover up of investigating the source of C19.

Labelling sensible hypotheses as “conspiracy theory”.

The Delegitimizing authority / anti-establishment / anti-woke nationalism and woke anti-nationalism.

Recent history last 50 years.

Because more people are online, the c19 pandemic has exasperated rifts, and general (inflammatory) woke dialogue fuelled and normalised the woke and cancel culture.

Since Socrates, our survival and effectiveness have been dependent on collaboration and good relationships. The key to success has been a common moral code – with collective responsibility and a focus on speaking openly and helping each other.

Over the last 50 years, the West decided not to decide all that we think morality should be, Thatcher suggested instead we should look after ourselves – leaving nothing holding us together.

The outcome of woke culture is putting oneself central within a set of tyrannical values and labels around the axes of woke culture principles - leading to power shaming, identity politics. This creates a concerted effort to obliterate our shared vision and family and individual identity and values. The selfish perspective and general dialogue escalate, provides (flawed) purpose and (flawed) meaning. The silencing causes an inability to get to the truth and fuels the next wave of propaganda creating menticide. This left tyrannies vulnerable and more manipulable to a takeover by rogue states such as China.

We are experiencing cultural climate change: In 1964 the use of "we" overtook the use of "I" in 1964 – more interested in ourselves. Lord Rabbi Sachs noted this led to what he described as cultural climate change. Issues like abortion – became legal – a new form of sovereignty of the individual?

Morality became what we chose it to be: The economic revolution 1980s: "me" not "us," social media – mid-1990s – self-marketing, 2000's taking selfies Facebook Instagram advertisements for myself, themselves. Curating self-images see and envy. Lack of concern or welfare for others.

The importance of a moral framework.

History suggests Woke culture inescapably ends up with genocide or racism, hatred, and violence without public conversation or a single moral culture.

We cannot leave a shared moral framework to the state. Morality means accepting responsibility, woke culture is about victimhood blame and labelling.

So Woke culture is a downward spiral that encourages nothing getting done.

We have moved to a shallow contractual model. A covenant is like a marriage vs contract vs one-off without a long-term partnership. Covenant is about respect and love for each other. American politics was built on a covenant – exile and stranger – the USA-based covenant, the most moral framework with a pledge of mutual loyalty and value of individual views and free speech.

Agreeing on values based upon common ground, secular grounds, has never led to a civilized or stable society. If you do not maintain values populations dissolve and people throughout history have disappeared.

Nothing else apart from religion has been shown to be a durable base for society's morality. Even the non-Christian Jordan Peterson suggests you cannot define your own purpose without a Godlike external influence.

Where did things go wrong in the 20th Century?

1918 post-war gave rise to a self-centred focus: the roaring 20s, wild dances.

This resulted in the 1929 crash, the 1930 depression, the rise of Nazi/fascism and WW2, the world was at war again.

The 1945 post-war period was the opposite moving to "we care for the less fortunate": the education act, secondary education for everyone in the UK (including the poor), NHS welfare state, USA GI bill – to help those who sacrificed in the war. The Marshall plan was graceful and provided loans to rebuild including Germany. This focus on “we” since has led to 75 years of peace.

Woke culture is an issue because if the focus is on “I” not “we” then we are in trouble. We need a strong society that no longer feels violent towards each other.

If not, history suggests we are and will experience increasing division anger and violence.

The sense of family was at the heart of those that risked their lives to save the Jews. These heroes came from strong family backgrounds, family is a source of moral strength.

Blaming other people is a sign of abandoned morality. Morality means accepting responsibility. A victimhood hierarchy is dangerous and equates to neo-Marxism.

A victim has lost the battle of their life, and victims handicap themselves for life.

Woke culture drags up failures of the past.

A mature culture talks about culture openly – what is wrong and acts collectively. We must do this as part of one community rather than allow pariahs to impose their views and silence contrary views.

Silence leads to violence… knocking down statues with racist and other accusations/labels. All you do is destroy history. Those people who disagree - this is what happened during Mao's cultural revolution? This is totalitarian we must reject this violence that woke culture is.

What is the woke dynamic?

We are experiencing the atomization of society. What will make decent people feel moved to act and resolve the silence?

A purpose-driven life: "it's not about you." This is what brings people together, we need people who disagree and sharpen axes we grow through arguing.

Disagreement, source of growth and truth and friendship – we should not settle for anything else.

Cancel culture happens at the CEO level, such as the CEO of Supertrends, relentlessly pushing for tech transformation, cancelling and not listening to those that do not fit. Where are the checks and balances in that?

The times when things got bad, Rabbi Sachs said, were when people looked after themselves rather than each other.

Redefined truths.

To have group thinking you need to have redefined truths; but who is qualified to set these redefined truths? Woke culture has very noble humble origins, slang in the US black community "woke to this" just meant to bring something to someone's attention. Since then, it has been hijacked by political correctness and anarchist anti-establishment elements BLM and others.

Woke dialogue is always centred around: race, sexuality, gender and feminism, status and power, social inequality, Offence hindering open debate – Snowflakes' views trump truth because they become offended. This fuels and encourages victimhood and arms the marginalized with a new purpose - they can feel moved into anti-establishment thinking rather than taking ownership and action.

The woke link with anti-establishment atomised society and technology.

There is a link between Crypto Currencies, Silicon Valley not paying taxes, Big Tech / social media controlling the dialogue, defunding the police, and China surveillance, etc.

The USA is becoming increasingly under the control of the Chinese: Chinese companies such as Alibaba are now USA-listed companies and Google is closely collaborating with a Chinese Giant.

The common link with tech is an interest in controlling free speech, society at the individual level, and anti-establishment sentiment. There is an unholy coalition appearing between some very unlikely places: China, Silicon Valley, Big Tech, and social media.

This raises big concerns about discernment – who decides what is right and wrong, who decides what is censored, or is it just defined by the offence taken by the victims of the system and the woke labels?

Woke untruths and out of control inflammatory escalation – the tools of silence.

“Black Lives Matter” is a clever exclusionary flawed generalization. This generalization creates a blind spot that negates the truth that in fact “all lives matter.”

People have been threatened with their jobs for suggesting the TRUTH that “all lives matter.”

The George Floyd killing murder was conducted by a group of police from a variety of ethnic backgrounds. The police deployed a standard Israeli security forces restraint technique they had been taught. Of the three charges the killer was convicted of, not one was for racism. Note the blind spot a white man was killed by the USA police the following week, yet the blind spot and fear of speaking out caused this white death. Note a white death does not fit the BLM mantra so doesn’t make the news.

This is how these woke mantras create a tyranny, or an inability to challenge or reach the facts, and a dangerous blind spot, and silence.

This woke culture is a catch 22, which leads to never getting to the truth - so no sacrifice or humility - and the scenario gets worse, and through engrained habitual thinking the fear escalates and behaviour becomes more tyrannical, because it is indefensible.

The escalating issue – the existential public observation of George dying in public, put people in fight or flight mode. This is when rational thinking goes out of the window and fear takes over fuelling the escalation of the tyranny and fulfilment of the “Black Lives Matter” flawed generalization rather than shutting it down to get perspective armed with the facts.

So, people cannot speak out; people lose the rational debate. Unintended consequences for those searching for the truth get fired or threatened with their jobs.

All kinds of generalizations ensued – accusations of systemic racism, fuelling victimhood and even anti-establishment neo-Marxist woke calls to defund the police.

Any negative generalizations are false including those that Woke culture is built upon.

The redefined truths are creating unintended consequences.

The endless complexity of Woke culture, Multi-gender LGBTQN..+ etc definitions or redefined gender labels. These redefined gender truths now usurp our biological sex. These are creating a contagion in young girls who empathise with their confused developing peers and in so doing enact the same behaviours. This for the first time ever is creating contagion and confusion. The stock woke answer is that the child's gender was always the truth, they were born that way, and this has just come to the surface and has been suppressed. This fits the victimhood of Woke culture.

This leaves doctors and parents unable to advise and help, for fear of being labelled as gender bullies, this has been enshrined in UK law. When in fact the endless LGBT labels themselves are the source of confusion and it's a natural process for children to be confused about sexuality and find themselves in their teenage years.

Because people cannot speak about the awful treatments for gender transformation, they go unchecked, and patients often regret taking that route.

This often ends in real misery when the issue of gender confusion is just part of a normal teenage transition.

Early indications that these labels are causing a spike in young female suicides as well as undermining normal heterosexual relationships, and the basis of the different male and female roles they play in a family and value in bringing up their children. Another subject that threatens your job.

Flawed mantras

The mantra of the dominant suppressing male patriarchy does not wash because dominant males do not survive very long in hierarchies if they don't put others first.

The equality mantra. The key distinction is in differentiating between equality of outcome vs equality of opportunity. For example, would you want to equalise the outcome in these two scenarios?

5% of women are in prisons should we convict another 45% of women to level this up? 1% of women are bricklayers. Should we force women to become bricklayers?

Low % minority groups in executive positions and “We must have black representation”, “whites don’t understand” – “colour defines diversity” these are blatantly a racist policy?

Martin Luther King put it perfectly “I look to a day when people will not be judged by the colour of their skin, but by the content of their character.”

Positive discrimination – based on race, is simply racist and negative discrimination for the excluded group.

Silencing the woke mantras do not allow for listening to the facts and debate.

The generalisation of “systemic racism” – a recent UK Government report found no evidence of systemic racism – yet people will not accept this, preferring to find facts to fit with their negative generalisations and accuse the reporters of being “unconsciously biased.”

This is notwithstanding that prejudice should be stamped out and opportunities should be open to everyone - equality of opportunity.

The link with technology.

The mantra of profit and relationships does not fit with the tech mantras of growth and domination, leaving tech companies becoming THE dominant intermediators of all sales putting them in control of sales and increasingly routes to market.

These tech systems do not fit well with our mind space for different views or breadth of thinking, because people cannot cope with too many opinions contrary to their own.

Our natural modus operandi is to surround ourselves with people like us.

Therefore, woke culture is attractive for people who feel alienated in some way and dangerously brings a new but flawed group and sense of purpose they may have lost.

The key paradigms to thrive NOW:

Labels kill dialogue – distort and delete the facts – which is literally dangerous.

These are the top 7 factors we have identified for businesses to challenge themselves to navigate. Notice how the blind spot is centred around more of the same tech, yet so much of the dialogue continues to be around digital transformation which has contributed to such dramatic reductions in relationships and profits.

Critical thinking "That don't make sense."

Facts vs philosophy. You can't rationalize philosophy because it's not directly based on facts. But you can rationalize the type of dialogue to get to a useful conversation.

Strategy is simple when you get to the facts, but philosophical mantras or labels do distort and delete the facts. This diversion takes the team's eye off the ball so flawed labels undermine the integrity of your people.

People with integrity and who care are rare. These people relentlessly value the truth and really value finding the truth and have the courage to call out lies.

Finding the truth is because of a relentless messy and uncomfortable conversation and a journey, rather than a destination.

There is far more we have no clue about than that we know or understand.

"The only true wisdom is in knowing you know nothing." ... Socrates

The trigger to challenge thinking and steer the ship is simple – "that does not seem right."

"We need the data" is a flawed mantra too. Waiting for the data is a convenient mantra too - a diversion/form of denial when the direction is obvious. Creating data is onerous to prove the obvious is hard work and a waste of time.

Anyone should be able to say, “STOP that don’t make sense”, which should trigger some critical thinking and discussion to steer the ship in the right direction.

“The value of thinking and who is steering the ship.”

If thoughts are unchallenged, they become facts in our minds. Open dialogue and critical thinking are the tools to challenge thoughts, philosophy and reach the facts.

The antithesis of this is to cancel or label those that disagree, as “racist”, “you aren’t x,y,z so don’t understand”. These are aggressive false woke labels that shut dialogue too and are emotive, subjective, and inflammatory.

It is ONLY the minority that complain.

Research suggests that 75% (and probably many more now) will knowingly do the wrong thing in organizations to fit in. With the threat of woke culture, the 25% with integrity are fearful of getting to the truth. The 75% are not thinking of looking for facts to fit with their mantras. In just doing so the truth makers become marginalized and immobilized.

If people do not care, we become manipulatable, and this reduces us to being a machine. Machines don't care.

Non-thinking, compromised, and manipulatable people are ideal for a tech or any totalitarian system.

The tragedy is that those that do care get bent out of shape and are at risk of getting incredibly stressed. This misalignment causes anger by forcing these important captains of the ship to misalign their actions with their beliefs.

The key is not to show that you are getting riled by the woke mobbing and watch and observe the communication. So, call out the nature of the dialogue, pitfalls, and traps, do not personalize the discussion or add labels that just fuel the fire.

It's not about being right or wrong either, it's about having the skills to engage in productive dialogue and maintain relationships and create empathy, and in so doing value DIVERSITY!

China.

China’s communist party uniquely does not care about people which fundamentally separates their strategy with the Christian basis of Western cultures, and maybe Muslim Arab countries too.

This leaves China's ability to experiment on artificial intelligence and intense surveillance unscathed. There is a parallel with this ethos and woke culture – group thinking and labelling trumping the value of individual thoughts. Thinking is not convenient for the dominant plans for the China Communist Party?

Developing anti-bias AI is ludicrous because AI depends upon bias.

A human being’s bias equates to their personality, view, and defines their diversity.

AI-driven weapons are fuelled by “if we don’t, they will" – this drives the development of these unethical weapons and the race to the bottom. Emotional intelligence has become far more important than IQ... Do you notice the flawed assumptions in this statement?

"President Joe Biden and his top diplomat have tried to say they want China to follow "the rules" while it pushes past the United States as the dominant power in the world?"

That is the only way to see the signals they are sending to the Beijing regime.

These are critical to recognize the blind spot, strategic misalignment, and inactivity.

Money is a funny thing; it can be printed and given to anyone; fiat currency structures are the guardians of wealth distribution. But unlike crypto, FIAT systems have a responsibility to protect national interests, wealth distribution, society, looking after the poor, infrastructure, and keeping the peace.

The woke should be concerned about taking away job opportunities from those with low IQ. They cannot be retrained or upskilled to do them!

A fiat currency's valuation is based upon basic economic principles such as profit, fair global trade, avoiding printing money, fair taxation, and interest rates.

This valuation is set by global markets relative to other currencies.

When it comes to cryptocurrency, crypto has none of these social responsibilities to pick up the pieces. So it appears attractive to elitist anti-establishment types or those that want to break away from the system and shed or just ignore these broader responsibilities

With so little of value to invest in, crypto appears attractive because it plays to the basic human flaws of scarcity, and greed - a potential get rich quick scheme.

So, investors leverage and pile on to manipulate the price ever upwards, but where is the value being created to warrant that value, it's purely manipulating the value of nothing?

Like tech intermediaries, cryptocurrency is taking cash and wealth out of the productive economy and out of the public sector fiscal wealth redistribution model.

Woke judgement and context in tech - there is a lot of doing being driven by the mantra of "digital disruption" but thinking is poor.

Investment in technology for the first time has overtaken the investment in the top investment – Oil.

Bitcoin is a textbook Ponzi scheme. The evidence is crystal clear.

This is an excerpt I shared to check the thinking: *"It has no intrinsic value. You can't eat it, wear it, or heat your house with it. Unlike gold — which at least feels nice and looks shiny on your spouse's ring finger — you can't even see Bitcoin. It is not a productive asset. It's not a factory that produces an item. It has zero underlying value. None. It's not backed by land or commodities or — as with national currencies like USD or GBP — the threat of violence (in the form of a wage garnishment, asset seizure, and imprisonment.) The price fluctuates so wildly (what healthy currency doubles in a month?), it's virtually ineffective as a safe representation of value or means of trade. Its value is solely derived from the trust that the price will continue to rise indefinitely. That there will always be new investors to buy out the old ones. Bitcoin is a Ponzi scheme…The major problem here is that most unsophisticated investors currently view Bitcoin as an investment. It's not — it's a currency, a vehicle of trade, a means to an end. Currency is the oil that keeps the engine running smoothly, but it's not the engine itself."*

The generalizations and labels in the responses illustrate the same type of dialogue. Note how they distort and delete the basics of the discussion:

"Classic example of "what man doesn't understand, he destroys"."

"And what you have done here Tom Pickering FIET is to lay bare your misunderstanding of crypto assets. "Unsophistication investors" such as JP Morgan, Goldman Sachs, the Norwegian Sovereign Fund, Michael Sayler, Elon Musk, Raoul Pal, Nike Novogratz, and many more."

"In college, we noticed all the easy classes had a lot of people wearing hats. In Bitcoinhaterland there's a lot of gray hair."

"OK, Boomer."

"Typical no coiner."

Notice the digital parallels and precedents – normalizing.

Normalizing shuts down fundamental thinking and debate. Here is a quote from a

very bright tech chap I know well... *"I get often asked why and how one would use crypto in the real world today. Let me use an analogy. When cars first emerged in the 19th century, they seemed like an absurdity. There were no roads for them. On dirt tracks their advantage over horses was non-existent. Many horse riders laughed about the silly idea that someone would ever choose a car over a horse.*

But roads started to get built. Slowly at first, then quickly. The absence of roads did not mean that a horse was a better medium to get from A to B than a car. Over time more roads were built, and cars started to challenge the horse more successfully... Make of that what you will."

From the Oxford Brookes Institute of Ethical AI

"Watch Professor Nigel Crook and Dr. Matthias Rolf's webinar 'Moral Robots and Ethical AI'."

Comment:

"Moral robots" and "Ethical AI" are misleading mantras that by association distort and delete the reality and shut the debate before you start. It is important to understand psychology and develop the skills or flawed strategies that will remain unchecked, and not pursued.

Getting back in control.

We must understand what is driving us and improve our own self-awareness by becoming aware of the nature of our own and the broader dialogue.

This is an EQ issue, not an IQ issue? EQ identifies the right issue to fix, and IQ is the ability to fix it.

Why is this dialogue and strategy out of control?

The people making these statements are not aware of the psychology of what drives them, such as: "if we don't, they will" and the lazy thinking and "racist" "you don't understand" labels to fuel the tyranny.

With stress, these labels become more attractive; they create certainty weirdly enough. Woke people have lost perspective and the art of healthy open dialogue and tech limits the mind space to have difficult discussions. It is maybe a good thing that our young are addicted to tech and too fearful too, else there would probably be an uprising.

The value of proper conversation.

Regardless of whether weapons or business, innovation using tech is nonsense too, we do not need it, we have got common sense and a proper conversation.

We need to develop the skills to STOP doing stupid things and focus on and develop the top 7 below.

People are not good at stopping doing things though. This requires un-learning.

So, by the time you have done something, even if it is wrong, there is a strong need to align your beliefs to the falsehood rather than overcome your pride to admit you are wrong and re-align.

Winningthinking® skills to get a grip of this.

Instead, we have woke culture which is the perfect group thinking/silencing tool that works perfectly alongside tech!

Watch out for negative generalizations, (they are always false), and listen out for people's assumptions that are driving the actions:

People do see woke culture as a big threat to wellbeing and humanity.

Technology thinking flaws and mantras.

Even with the fundamental flaws in technology, the overwhelming mantras are not causing people to challenge the fundamental applications and suitability of tech. It is obsessive, and the debate is wholly technology-dominated, inhuman, and narrow.

Here are some bizarre generalizations that imply and by association create blind spots from the House of Lords All-Party Committee agenda:

"How has happiness and wellbeing become intrinsically linked to our digital lives?

Can happiness exist in a workplace without purpose, fulfilment, accountability, or ownership?

How are art and music influential to happiness?

Where does happiness reside at the workplace? How can homes, businesses, and society use technology to keep people as happy as they can be? "

The Bitcoin technology: “more tech will fix it”.

The only way to ensure you get 50 matches in a box is to put them in the box in the first place. You cannot "count in" quality. In the same way, Crypto tech needs fundamentally challenging whether it is the right solution. If the outcomes are bizarre there is something wrong and needs rethinking:

“#bitcoin is the ninth most valuable asset in the world, now requires nearly as much energy as the entire country of Argentina. The Bank of England had warned that cryptocurrency had “no intrinsic value” and claimed investors would lose supporters’ money. The Sierra Club study pointed out that if Bitcoin mining processes carry on the way they are, in 30 years' time they could increase the world's temperatures by 3.6 degrees Fahrenheit (two degrees Celsius). The environmental impact would be catastrophic.

AND “Bitcoin’s energy consumption has more than quadrupled since the beginning of its last peak in 2017 and it is set to get worse because energy inefficiency is built into bitcoin’s DNA.” Charles Hoskinson

BUT... The cryptocurrency had previously hit record values with both Ethereum (ether) and Cardano (ADA) hitting all-time price highs.

Good judgement.

The trigger for critical thinking is simple "that does not seem right."

Good judgment is an exceedingly rare trait and extraordinarily difficult to measure.

"I’ve met enough #bitcoin enthusiasts to know one thing: These are not the people I want as stewards of a new financial era." Source Medium

There are fundamental issues with bitcoin for example excessive power consumption is a factor of an escalating cyber risk which could bring the whole cryptocurrency system down with no accountability to investors whatsoever. So, worthy of some critical thinking?

Flawed mantras – “Endless Growth”.

You cannot move forwards without sacrifice. "Endless growth" creates accumulating lies, and right now specifically these lies are contributing to some sort of (tech) or woke hell.

So, the flawed solutions create a false and very damaging asymptote - so the outcome will never be realized, exasperating mental health.

It is important to identify and become aware of the mantras that create and fuel these corruptions and tyrannical outcomes.

Correcting the misalignment.

It is important to subject ideas that are wrong to open scrutiny and free discussion without fear.

You need new skills to get to grips with this corruption and break these tragic catch22 outcomes and relentlessly search out the truth. So, is being unstoppable what good looks like?

Mantras cause actions and the challenge becomes misaligned. This is the cause of a company turnaround, but, like not noticing your hair growing through familiarity, these issues only become clear when things get tough. Or many businesses carry on profitably regardless, even though they could do much better. winningthinking® shows you why this happens and what to do to prevent this and how to become unstoppable and stay on the truth.

The Winningthinking® skills share what we have learned having resolved 200 company turnarounds – difficult circumstances are the acid test.

Breaking through the fog.

Maybe now you can see how the same behaviours are shown and conveniently used by woke advocates to operate in the fog and fuel of untruth. As part of the winningthinking® program, we share case studies that explain why and share why our leadership skills and even basic conversational skills are defunct. It is important to look after your best, highest integrity people; these people are those with the courage to stand up and get to the facts, so the right people thrive.

Operating in the fog of untruth is not a leadership trait you should encourage; it encourages the wrong people to thrive.

When you get to the facts – then strategy is easy.

The challenge is getting to the truth and staying there. How can you do this in a sea of silence when the good people are silent, too afraid to speak up?

Winningthinking® gives you the opportunity to take different actions from which I hope you will find yourself and become free and unstoppable.

This also helps break through glass ceilings when the EQ or thinking is the limiting factor and there is IQ headroom to build upon.

Gaslighting can be subtle, inadvertent, and deceptive.

Gaslighting is a form of psychological abuse where a person or group makes someone question their sanity, perception of reality, or memories. Here are some examples:

Context: US Christian teacher suspended for refusing to say that a 'biological boy can be a girl, and vice versa' - Premier Christian News.

Responses to the post illustrate some of the blind spots and gaslighting or those that have been "gaslit":

Complete dismissal, playing the taking offence card:

"How completely ignorant Tom. Let's just start with the basics. You are so off-base it is sad. Would not make any difference if your words and actions did not hurt others, but alas, you actively do hurt others." Educationalist

Comment: Offence card to marginalising traditional sources of morality/ discernment, *"well they can... I think it is probably for the best to get religion out of schools and rely on science, just an opinion"*

Comment: dismissing religion

"I have another view on Woke in that it's about correcting Inequality and Racism."

"I listened to this, but still believe that Woke culture general means Anti-Discrimination - which must be a good thing. If you were BAME or Disabled, you would understand about everyday discrimination."

Comment: Duped by the mantras and you do not understand

Context from a different perspective: what is a white male view of woke culture?

"Tom, I'm glad that I'm only seeing this now. I need not say more as Asif has made his points brilliantly. What are you trying to prove? It's really sad that you can't see that your own bias towards woke culture is getting in the way of you having a true and believable survey and argument without people thinking you are being provocative and attention-seeking."

Dismissal of a different perspective and waiting for a bear trap......

The voting reflects your narrative in the posts you have responded to on this thread. In the future, I'd suggest you wait till the survey has been completed before deciding to skew your own results and influence how people should vote!

Your poll is now null and void. To be honest that's what you wanted anyway right?

An audience to share your uninformed views without really doing any serious research or truly understanding what woke means. The term has been around since the '60s (I would suggest you start your research here)! The term is meant to denote an attentiveness to important facts and issues especially issues of racial and social justice. Why would you have a problem with issues that many allies and people of color are voicing around societal injustices? I'll leave it there Tom as i actually think your heart is in the right place, but your arrogance, tone, and tunnelled vision lets you down. Diversity Chief NASDAQ quoted Co

Comment: direct personal attack - labelling as racist, etc:

“Are you living in Apartheid South Africa White only voting? Not facing the issues/ obfuscation / you don’t understand” “you have phrased the survey in a crude and biased way. Limiting views as can be seen to virtually a minor audience and as such your overplaying your personal views. I deal with a lot of company owner's executives and frankly, not a single one has raised white make woke culture as an issue. They have raised rudeness and overbearing ways white males and Asian males have behaved. Your views are not a major issue with businesses. If your small enclave of white males has an issue they should go on courses and learn how to manage and lead and encourage positive critique and be ready to take it. Wise up to the young multicultural world.

Perhaps you should spend time on non-white makes experiences.”

Comment: label, run, no interest in engaging in a discussion: “Opinions based on fact have value, Tom. Opinions based on ignorance have no value. “

Comment: hi, thanks, specifically what are you referring to? Hit and run! Silence – no interest in engaging in the discussion.

The implications of the tyranny.

The woke outcome is to weaken and undermine the ethical fundamentals of society, and faith, with redefined truths that allow no alternative views. The outcomes are far-reaching in sport, redefining transgender athletes, gender confusion in schools, with a contagion impact on girls it would not be so bad if the dialogue encouraged diverse views. These are rigid tyrannical mantras that are not based on biological fact, for example male and female, or mother and father.

The category missing from the LGBT etc is the majority - heterosexual.

The woke "patriarchy male mantra", is nonsense because organizations simply do not hold together if males are overbearing.

A neo-Marxist style with labels that lead to conflict, division, anger, and silence, undermining a perfectly reasonable diverse set of beliefs.

The response is that "you don't understand".

We need anti-bias retraining. It is more closely akin to Uyghur concentration camps in China than diversity, it is an assault on society.

Being forced to believe something is extremely unhealthy particularly when it is based on redefined truths, untruths, or anti-truths, if not out and out lies.

Relentlessly looking for the negative, taking down statues denies progress, and is more akin to the cultural revolution in Maoist China.

Woke is a tyranny because people are getting fired for failing to support the bizarre notion that "a girl could be a boy", or "the importance of the different unique male and female roles in bringing up children" (the norm).

The accusation of being exclusive, racist, and undermining male and female roles undermines the basic family unit that holds the morality of society together.

The dialogue is totally humourless, aggressive, and denies any beauty of any kind.

The danger of tyranny is that you can't make something wrong – right, you must sacrifice something to progress, and without open dialogue you are stuck.

Being forced how to think.

Would you prefer to be forced to think that these woke mantras were true or forced to believe in and follow the Chinese Communist Party ideals, whilst being under intense surveillance? Maybe the latter at least China has some purpose and ambition!

China has taken an unusual twist from woke culture by focussing on capitalism to improve the wealth and fuel the domination of their plans, whereas North Korea killed enterprise and used starvation as a key tool to control their population. The fear of your peers "informing the authorities" is now as prevalent in the UK as it is in China or North Korea.

Because the false woke mantras are self-fulfilling, ethical standards deteriorate, and control increases because people become punch drunk - silence corrupts and stops people speaking out. Only today, abortion legislation at any time during a 9-month pregnancy was defined as a human right, proposed by the EUC.

Facebook added an emoji for pride month – fuelling this echo chamber and tyranny and their control because woke culture is a convenient tool to silence and control their critics, without scrutiny nor fear of reprisal.

The top-level mantras of victimhood, diversity, offence, sexism, transphobic, and systemic racist accusations are the basis of the gaslighting and labelling of critics, and tragically are being enshrined in law without a proper definition of either what constitutes "racism" or what constitutes giving "offence".

Whilst China covers up C19 and Uyghur's scenario China gets things done by silencing their critics with an element of courage and stoicism, whereas woke culture in the western form creates a toxic mix of stasis, lies, and control that bends the best people seriously out of shape.

The tragedy remains in that the flawed generalizations are a comfortable modus operandi and create (flawed) virtuous purpose. These generalizations value and are built upon victimhood and blame and are fuelled and kept alive by endless criticism. This operandi is the antithesis of taking personal responsibility for oneself and acting with the grace that being aware of our own flaws rightly warrants.

Silence and defending the obvious.

The consequences of not healthily challenging the facts and preventing open debate are that these woke thoughts become facts and keep religion (the traditional and only proven moral codes) out of business and schools. This marginalizes the only moral codes that have held societies together for millennia.

These familiar religious or male and female family moral codes are difficult to defend because they are so familiar, so we become blind to the logic because they do not need to be fundamentally challenged.

Woke culture just gets worse and more controlling because there is no sacrifice and freedom of speech gets reduced – which undermines good people's integrity forcing them into silence.

Practical new perspectives to consider.

You do not know better and nor do I.

Acknowledge and love the other person for who they are.

Take an interest in understanding a different view, and where they are coming from and accept that is OK.

Use humour, do not take yourself too seriously: humour levels assumptions to enable discussion so breaks the deadlock.

The purpose of dialogue is not to fuel your ego, or not being right or wrong, it is about building the winning thinking skills to be able to have productive conversations and dialogue to inform decisions and take the right action.

Identify this, what is stopping that person from taking ownership of themselves?

Trust people to have their own moral compass/rationale. Respecting individual uniqueness and diversity. Discover yourself by aligning actions with your beliefs, give people space and opportunity to take their own new actions.

The freedom to act aligns actions and beliefs allows people to find themselves and this creates grit. If there is no new thought or action – no grit.

What prevents progress?

Do you read content that you disagree with? What does your dialogue look like if you speak to someone you disagree with?

The issue to address is becoming aware of the difference between EQ vs IQ. EQ identifies the issue to fix, IQ is the ability to fix it. The EQ-related dialogue distorts and deletes the actions vs the target. New skills / winningthinking® game-changer program and tools required to raise the cause of this and fix that.

A summary of the key tools.

Become aware:

Avoid using generalizations - they create confrontational dialogue.

Labelling creates roadblocks and fuels and escalates the conflict.

Do not add YOUR meaning to what was NOT said.

Avoid taking comfort in herd responses, this leads to mobbing – bullying individuals.

Do not respond to generalizations, get to the specifics – what specifically do you mean?

Get to and face the specifics of concern then you can break the conflict and get to a productive dialogue,

Stick with these tools - this is not easy and still requires courage.

Practice the perspective of and value humility.

When we are weak, we are strong.

Practice, enjoy and relish the personal development opportunity for your own development and enhanced relationship with another by welcoming the challenge, taking an interest in understanding where someone is coming from, and valuing their different view. They care!

Good relationships are based upon challenge and are strong enough to allow for challenging each other.

You do not know what is right you just have your view.

People are always doing the best they can.

Reject your ego or your focus on always and relentlessly being right, rather value dialogue/quality of conversation and understanding of another's point of view before commenting.

It is still ok to disagree, acknowledge it and move on.

Start the discussion when you understand another person's point fully – a good place to start a discussion.

Distortion of actions

Accept and understand if you do not address and challenge your thinking or mantras, your actions will get distorted, and facts deleted.

For example, The Paris Climate Change agreement if implemented in full claims to only impacts 0.25C of the inevitable 0.4C by the end of the century. Yet it is perceived like an existential threat and only 1 factor of many.

Become aware of what makes you angry, do you find purpose in the woke mantras and comfort? If so, you will experience a personal existential threat if someone challenges your views on woke culture. Woke is powerful - could be perceived as a religion.

Avoid debates if you are angry: the BLM protests were perceived as an existential threat following the shocking murder of George Floyd.

This creates distortion and like climate change an overreaction.

Stay busy. Have we had it too easy for too long so are finding fault when we should value and be grateful for what we have?

The human flaw of certainty.

Generalizations flourish in uncertain times

Recognize the convenient and flawed comfort found in generalizations.

Generalizations are comfortable because even though generalizations are flawed, they create certainty. People crave certainty in particularly uncertain times and will do anything to reach certainty or a "return to normal".

Human beings will make bad decisions to reach certainty.

Focus on the other person - giving is 50x times more pleasurable than receiving and it is a curse not to appreciate what you already have.

The related obsessions and parallels with technology and mantras.

Corruption by investors making these flawed choices and through pride fuelling valuations trying to make them “right.”

Company turnaround and crisis are caused by generalizations and mantras distorting and deleting.

Carillion collapse was caused by the mantra “focus on cash before profits” which drove their funding model, and probably corrupted auditors causing an inability to speak up.

Or “We can’t find the right people” causes micromanagement causes no decisions or implementation.

The mantra of endless growth.

Burning assets/resources recklessly – profit and growth. Businesses are far more sophisticated to run now – unwind the mess.

Yet listen to the mantras - still focussed growth and domination – accelerated by tech regardless of profit. Profit and relationships should be front and centre – yet businesses are still obsessed with growth and domination.

Why can't we stop PR machine?

We are not thinking, value thinking, how can you tell if someone is thinking?

Generalizations prevent thinking and block the prompt / trigger for critical thinking, doing vs thinking, new action.

Where do you find your discernment – do you define that yourself or do you have faith? What steers you or your conscience?

Be humble we will never understand much of what surrounds us.

Socrates "The more I know the less I realize I know."

Open the dialogue.

You can discover the blind spots by reading, valuing, and opening the dialogue with your peers, to continually steer your actions and ship in the right direction.

Learn from others, what thoughts come to mind when you think about reading a book?

What can I learn?

Why might this person know better than me?

Notice the blind spot: I know better?

Challenge your thinking.

Do you search out beliefs that fit with your thinking or reinforce your own beliefs?

Fear: Acknowledge when you are fearful or uncertain, this is a time you will shut alternatives out and you will do everything to get back to normal.

Brexit vs non-Brexit debate was based on small differences of view yet watch polarisation and how the generalization created conflict.

We cannot cope with the volume of communications let alone the multitude of different views; we tend to surround ourselves with people like us.

Become aware of the aspects and outcomes that typify fear:

Polarisation

Generals – conflict

Fear of speaking up - silence

Blind spots

Tech claims.

Tech claims to transfer power to the individual yet radical or innovative views are marginalized, leading to manipulation. Echo chambers, distortion, and deletion.

Fear of shame that comes from online bullying creates tyranny so cannot speak against it, yet social media is becoming the main form of communication.

Problematic responses.

Groupthink/conformity

Obsession with finding data to inform decisions when faced with obvious stupidity.

Or more “More technology will fix it”.

What are your and their underlying assumptions that you have not discussed?

What are your organization's related mantras and how do they fuel a nonsense strategy?

“People are machines?”

Augmented intelligence: “We can fix that with more tech”.

The implications of ill-thought-out strategy and blind spots are significant, so are creating impossible issues to fix such as opening wide to cybercrime which will be dominated by China’s domination of Ai and quantum computing. Or statements like “We don’t need to trust in supplier relationships – blockchain sorts that out” or “Technology is the future” House of Lords APPG.

Profitless growth models – it is important to make money – it gives and buys you time to think.

Is there any value in your service if people are unwilling to pay for it?

Challenge of the woke pillars or truths.

Become aware of your redefined truths and their dangers, understand alternative and traditional views.

Acknowledge the difference between IQ vs EQ: EQ enables you to fix the right issue, and IQ enables you to fix the issue.

Are you playing the player or the ball?

What does good look like, and what is a route to getting there?

Take NEW action from these perspectives:

Value teams and discussion.

Feedback is the means to steer your actions.

Become aware of the perils of having to fit in with values.

We are all so unique.

Develop the self-awareness of your own language and that of others by becoming aware of the mantras that drive you and what is driving others.

Become aware of how your mantras or generalizations create blind spots.

So, if you “Don’t like Trump” ask “what is good about Trump?”.

Take a great view of the nature of people, they have the infinite capability.

An individual’s limitations are defined by the opportunity/environment you provide as a leader, not them.

Strategic realignment.

Staying on track – healthy open dialogue as above.

Encourage everyone to think and take ownership of fixing problems in their own way.

Get everyone thinking and asking questions and leave ownership with them to completion.

See the value in taking new action.

Become aware of and recognize victim mentality – symptomized by blaming others or external factors – take ownership of what you can do and act now.

Create your own jazz band.

Value intimacy and informal discussion to develop empathy.

Identify and call out the types of dialogue to yourself and peers to improve your communications.

Do not focus on you being right or wrong.

Fearlessly in your own way speaking up about anything that does not seem right and does not back down.

Critical thinking triggers, “that does not seem right,” find out what do they mean?

Forgiveness and contrition are the basis of trust, not doing what you say.

Give people grace and avoid self-righteousness – the antithesis is judgment and blame.

Drive for a breakthrough – identify the flawed mantras causing roadblocks, distortions or blind spots.

Negative generalizations are always flawed, ask what specifically...

Become aware of the impact of generalizations, good or bad.

Challenge and hold each other to account.

Breakthrough requires new action, new action, requires new perspectives.

Next steps – winningthinking® application as a team.

Develop intimacy and empathy - practice the top 12 winningthinking® skills and perspectives, take the – winningthinking® game-changer program as a team – proactively take new action and realize what good looks like, open the dialogue.

This steers your ship in the right direction and wakes you to what matters. And keep the great black slang fresh by continuing to challenge your peers with love for each other.

Let us not deny or reject the emerging challenge…

If you are taking offence, challenge yourself – what is your flawed un-truth that needs challenging?

#Wokecheck... I spoke to a chap who trained a UK female Olympic cyclist yesterday. He said that she was refusing to enter the Olympics because the males that define themselves as females have made nonsense of the competition and made it plainly unfair.

Get out of the fog: how can these gender untruths be allowed to become so predominant and be allowed to corrupt the hard truth and biology that defines the male or female sex?

"Athletic teams or sports designated for females, women, or girls may not be open to students of the male sex," states SB 1028.

"For purposes of this section, a statement of a student's biological sex on the student's official birth certificate is considered to have correctly stated the student's biological sex at birth if the statement was filed at or near the time of the student's birth." Recently, several states have considered, and some have enacted, legislation aimed at preventing biological males from competing in sports designed for biological females.

What does good look like?

Simply acknowledge people for who they are, who they are taking an interest in and valuing different opinions.

The crux of business is profit and relationships, making money is NOT a rude statement.

Challenge yourself if you find you are blaming others or being spiteful you will be taking a victim mode of being and making things worse. Become aware of your own language, stop blaming, take responsibility for yourself.

You know perfectly well where you are kidding yourself in your language or contributing to convenient lies. These will be the things you do not like to discuss or make you angry.

Regardless of your colour, race, gender, sex, etc, there is nothing stopping you from taking control of your life. Listen to your own dialogue and become aware of those things that you know you need to do but are avoiding and find the courage to face them and resolve them now. Speaking your being and acting in this way has real promise, defines your purpose, energizes, and removes your limits. It is down to you. Lastly, be vulnerable, do not constantly pathologize society that is hell, take ownership of yourself, take an interest in others' views, and live forthrightly in truth.

Thank you, please share your thoughts.

Tom

tom@ewhurstengineering.uk

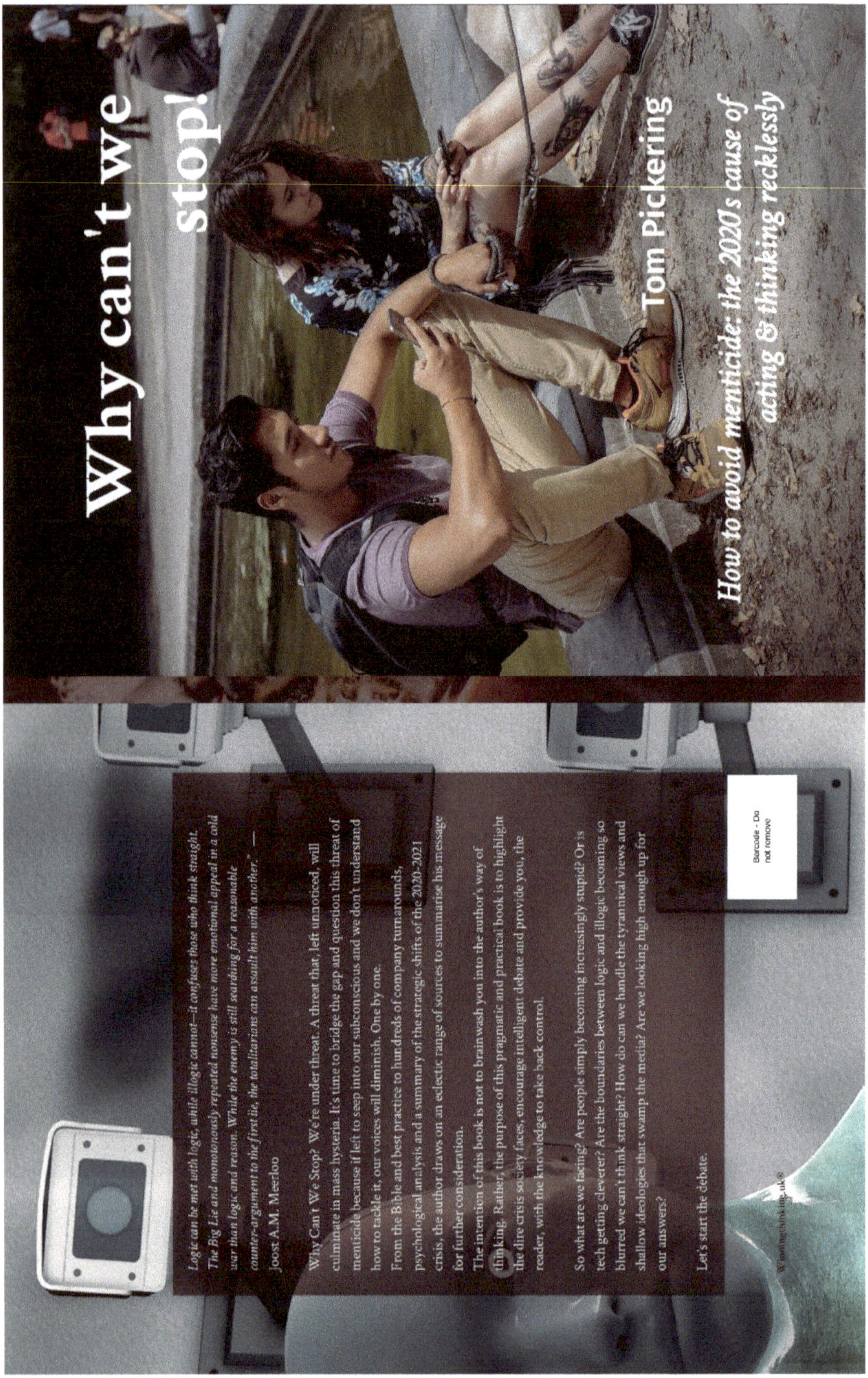
Logic can be met with logic, while illogic cannot—it confuses those who think straight. The Big Lie and monotonously repeated nonsense have more emotional appeal in a cold war than logic and reason. While the enemy is still searching for a reasonable counter-argument to the first lie, the totalitarians can assault him with another." — Joost A.M. Meerloo
Why Can't We Stop? We're under threat. A threat that, left unnoticed, will culminate in mass hysteria. It's time to bridge the gap and question this threat of menticide because if left to seep into our subconscious and we don't understand how to tackle it, our voices will diminish. One by one.
From the Bible and best practice to hundreds of company turnarounds, psychological analysis and a summary of the strategic shifts of the 2020-2021 crisis, the author draws on an eclectic range of sources to summarise his message for further consideration.
The intention of this book is not to brainwash you into the author's way of thinking. Rather, the purpose of this pragmatic and practical book is to highlight the dire crisis society faces, encourage intelligent debate and provide you, the reader, with the knowledge to take back control.
So what are we facing? Are people simply becoming increasingly stupid? Or is tech getting cleverer? Are the boundaries between logic and illogic becoming so blurred we can't think straight? How do can we handle the tyrannical views and shallow ideologies that swamp the media? Are we looking high enough up for our answers?
Let's start the debate.
Barcode - Do not remove
Winningthinking.uk®
Why can't we stop!
Tom Pickering
How to avoid menticide: the 2020's cause of acting & thinking recklessly

Why can't we stop introduction.

How to avoid menticide: the 2020's cause of acting & thinking recklessly

Tom Pickering

Winningthinking®

“The privilege of a lifetime is to become who you truly are.” Carl Jung

How to avoid menticide: the 2020's cause of acting & thinking recklessly

This is the know-how to free yourself, become aware of, and have the skills to avoid menticide.

Menticide is the 2020’s systemic Tech and the cultural threat that is destroying your humanity and your mind.

It took a few strong people to unravel the terror of the Gulags in Russia, not the masses. That might be you?

Whilst technology is accelerating this demise, menticide is nothing new. For example, in the UK during the 1500s, **menticide** justified burning at the stake as a form of execution for women found guilty of witchcraft. Most accusations of witchcraft did not originate in the church but resulted from local personal rivalries and disputes.

We're under threat. A threat that, left unnoticed, will culminate in mass hysteria.

It's time to bridge the gap and question this threat of menticide because if left to seep into our subconscious and we don't understand how to tackle it, our voices will diminish. One by one.

From the Bible and best practice to hundreds of company turnarounds, psychological analysis and a summary of the strategic shifts of the 2020-2021 crisis, the author draws on an eclectic range of sources to summarise his message for further consideration.

The intention of this book is not to brainwash you into the author's way of thinking. Rather, the purpose of this pragmatic and practical book is to highlight the dire crisis society faces, encourage intelligent debate and provide you, the reader, with the knowledge to take back control.

So what are we facing? Are people simply becoming increasingly stupid? Or is tech getting cleverer?

Are the boundaries between logic and illogic becoming so blurred we can't think straight?

How do can we handle the tyrannical views and shallow ideologies that swamp the media? Are we looking high enough up for our answers?

Let's start the debate.

More recently very few saw the rise of the Nazi party either and when they did, no one listened, and it was too late.

"He who dictates and formulates the words and phrases we use, he who is master of the press and radio, is master of the mind."—Joost A.M. Meerloo, *The Rape of the Mind: The Psychology of Thought Control, Menticide, and Brainwashing*

It feels weird to watch a car crash in slow motion that so few can see or say anything about?

The importance of this philosophical question is becoming critical to our human and business survival on every front, environment, sustainability, wellbeing, profits, relationships, peace, dialogue, etc:

1. "Do we have the courage to recognise a God?"

2. "Do we think we can set our own purpose and values i.e., know better?"

3. "What do we love and or who / what do we worship?"

Atheist secularism seems to be creating an endless list of tyrannical low-level shallow ideologies, with endless unintended consequences. Technology is recklessly accelerating and fuelling the process of

inflaming dialogue. Online communication does not accommodate different views.

So, are we looking high enough?

This book explores some of the consequences and opens up this debate and solutions.

The good news is that this threat and the unique set of winningthinking® skills my team has created enables us to be uniquely and peculiarly familiar with the challenge and there is a simple solution to it. That said, I am primarily a strategist, an executive operator, and doer.

But you do face an unprecedented threat engulfing society that few recognise, understand, and at worse can't escape – menticide, broadly the destroying of your mind.

"Where thinking is isolated without free exchange with other minds and can no longer expand, delusion may follow. Whenever ideas are compartmentalized, behind and between curtains, the process of continual alert confrontation of facts and reality is hampered. The system freezes, becomes rigid, and dies of delusion."—Joost A.M. Meerloo

I observe, and I am concerned you are facing a threat you don't see, and increasingly unless you act against the threat, your ability to influence, maintain your awareness, and capability to cope with the scenario, now and in the future, will exponentially diminish and the risks will exponentially escalate too!

This is a pragmatic summary checked from the broadest perspectives, drawn from hundreds of company turnarounds, observations on best practices, detailed psychological analysis, and strategic shifts drawn from the 2020-2021 period.

At its simplest level, I explore why we live in an environment that no longer meets our human needs and is one we can't make sense of.

The good news is that the solution can be achieved in a day, is remarkably simple, yet difficult. But most leaders don't see the need or connection.

The key capability in organizations is to:

1. Create a great place to work that: values individuals and lets them do their jobs in their own way

2. Seamlessly adjust to the target and strategy as it shifts

3. Take new action, and understand what is preventing new action

4. Work in a way that builds capabilities, relationships, and profits on the job

As the specialist winningthinking®, we capture this using the jazz band metaphor.

However, there are forces at play subconsciously usurping your agenda, the mechanism which needs to be understood to recognise their traits and mechanisms.

1. *The Evil of Silence: Woke Culture And The Mechanics Of Tyranny*

2. *The Psychology & Pathology of BigTech: Technology is now the dominant parasite and the psychological pathology that has reduced you & your business to dope on a rope*

3. *The 2020s The Decade of Menticide? Your Choice. What is the link between burning witches at the stake, BigTech, and how to avoid the destroying of your humanity and mind?*

My motivation is captured in my last book that "You have infinite capability: and you can regain what good looks like. But until you understand the three, 1-3 above, intertwined forces trying to get control of yourself and your business and dupe you off your agenda, you have no chance of getting ahead."

The exciting part of this is that the One-day team winningthinking® game-changer strategy workshops we run on the Ashridge Business School Campus can resolve all these issues in one day.

Whilst you really need to walk through the process with your team together in the context of your business objectives, I hope these books provide new perspectives that trigger some fresh thinking to enable you to transform your business.

The question is do you realise what is driving you and stopping you from achieving your desired personal and business outcomes for you and your peers?

If you have any questions or observations feel free to reach out on email tom@ewhurstengineering.uk.

PS. This book gets back to basics, so is even checked from a pragmatic biblical perspective, because it's quite clear that even people's own ideology is insufficient and is causing substantial unintended consequences at scale, so ideology and low-level fixes are not fixing the issues.

Why do we live in a culture that doesn't now meet our needs?

Are people getting increasingly stupid, whilst tech is getting cleverer?

Do we want these tech outcomes? Why do we think these tech outcomes are inevitable?

What is the source of wisdom in your organisation?

Why should we completely revisit what is right or wrong in business?

The infinite element is through a connection with God or else your limit is you!

A real crisis such as Covid 19 sorts out what and who is important and it's crucial not to ignore and apply the learning. Ignoring this leaves an internal misalignment which can be like an itch you won't scratch and becomes very stressful.

It seems very clear that with the level of dysfunctionality, anger, and unintended societal, environmental consequences taking place right now, we have been found out. So, we are unable to set our own ethics, purpose, and values in business.

Do you think you have sufficient wisdom to navigate your current scenario, and know where to turn to when you struggle?

Society and businesses are facing an epidemic of smearing individuals without debate. Debate is tough and the sign of wisdom requires people to understand that wisdom comes from something higher. Considering that we might be that source of wisdom is an interesting territory, which works until you face a tough challenge. Is it easier to avoid debate than resolve tough issues?

Even good secular attempts, driven by the smartest people, to resolve issues such as the Project Oxygen at Google are reported to have unwound. Even the best secular efforts to close the leadership gap without having a satisfactory source of wisdom, creates an endless stream of unintended consequences and misalignment:

• 10% of the directors promoted to a vice-president level position were recipients of Google's Great Manager Award.

• It is not enough to highlight the high scores, especially in a working environment that produces game-changing applications like Google. (Savic & Shi, 2010).

So secular institutions have no implicit wisdom.

Our intelligence is determined by our ability to fix problems which are largely determined by our relationships with those around us., and whether we fix the right problems depends upon whether our thinking is being challenged or checking if our “truths” are true.

So, if technology erodes these bonds between us and our relationships and quality of dialogue, are we are starting to behave stupidly like machines? Is this machine-like behaviour evidenced by our increasingly binary or polarised views and the inability to debate, so we start behaving like "non-thinking," just "doing" machines?

If this is the case in an increasingly technological society, the machines sure will win.

People do want to put humanity at the centre of a business, but do we understand what it is to be human?

So how do we regain morality?

It wasn't the majority that unwound and exposed the Gulags, it was a few incredibly strong individuals.

So, we must re-engage the minority that cares, and let them take a lead again and listen to those that we don't agree with?

But from where do we derive our wisdom to stand up against what we believe is wrong?

Mass hysteria or mass psychosis is an extreme form of groupthink when members of the group follow the consensus instead of thinking critically and valuing the truth and healthy open debate to inform decisions.

George Orwell said, “in a time of universal deceit, telling the truth is a revolutionary act!”

“What is needed now is a swift cold shower of reality. The time for self-harming distractions with "wokus pocus," obsessing about sexual dysmorphia, Marxist "critical race theory" and "climate catastrophism" ... - all of which the Chinese Communist Party is glad to encourage - must end.” GateStone Institute 19/8/21

Consider this: What are you too afraid to discuss that you know needs challenging and thinking through?

Over the last 40 years since the 1970s, there has been a switch away from frugality and common sense to "more." This has evolved into the mantras of growth and domination and the hyper valuation of tech companies.

These mantras have been found out during the Covid-19 pandemic with a shift and realisation of the importance of relationships and profit. The crisis has shone a light on who the heroes are that is the nurses and not leaders or social media influencers.

Yet the growth and domination mantras continue to be fuelled by a tech engulfed consumption growth-driven society, and social media, and a loss of purpose.

"Man cannot stand a meaningless life. The least of things with a meaning is worth more in life than the greatest of things without it." Carl Jung

"The only meaningful life is a life that strives for the individual realization — absolute and unconditional — of its own law. To the extent that a man is untrue to the law of his being, he has failed to realize his own life's meaning." Carl Jung

So, a significant wellbeing gap in businesses has opened, highlighting increasingly the meaningless purpose of many businesses, as well as a non-compelling proposition, particularly businesses centred around tech.

How we have lost control of technology is raised in The Pathology and Psychology of Big Tech.

Leaders / big4 are pushing more tech and driving the propaganda to influence the mass thinking to a tech end as if tech is inevitable.

So, do we need to start doing meaningful things and realign our businesses accordingly?

These are the top 7 strategic learnings we have drawn from the 2020–2021 period.

Getting in control of technology

Whilst the issue with technology is a psychological one, causing this mass psychosis, the notion that technology will self-correct, and laws realign to meet our needs, is probably flawed.

Why? Because technology is unlike humans. Tech is tireless, moves too fast, does not follow our logic, and impact at a time when, until now, we have never really needed to understand humanity, let alone the psychological impact of this deluge of technology on the basics of humanity and life.

Whilst Hitler or any other tyrant eventually got found out and shamed or usurped by the masses propagating their demise, a demise of technology is unlikely because in a tech-fuelled society our ability to organise ourselves and have meaningful debate across divides is rapidly diminishing. Conflict is increasing. Free speech is the means to correct and make sense of the challenge, not tech!

I think it follows that the emergence of binary thinking and polarisation and echo chambers are akin to people beginning to imitate technology, thereby becoming more like machines, and we are becoming far less interested in others' views or valuing people for who they are.

What are the underlying interests and drivers of technology?

I think it's accurate to say current tech, purpose, values, and cultural strategies are encouraging mass psychosis, which is the bedrock of menticide.

Mass psychogenic or mass hysteria is the rapid spread of illness affecting members of a cohesive group.

Mass hysteria was the cause of burning witches in the Middle Ages. The tools like cancel culture are akin to victimhood, scapegoating, and punishment that underly technology-centric solutions tools. Technology by its nature is binary, so doesn't accommodate free speech or free-thinking.

The process of menticide is common to tech or other tyrannies:

"There is another important weapon the totalitarians use in their campaign to frighten the world into submission. This is the weapon of psychological shock. Hitler kept his enemies in a state of constant confusion and diplomatic upheaval. They never knew what this unpredictable madman was going to do next. Hitler was never logical because he knew that that was what he was expected to be. Logic can be met with logic, while illogic cannot—it confuses those who think straight. The Big Lie and monotonously repeated nonsense have more emotional appeal in a cold war than logic and reason. While the enemy is still searching for a reasonable counterargument to the first lie, the totalitarians can assault him with another."

—Joost A.M.Meerloo, *The Rape of the Mind: The Psychology of Thought Control,Menticide, and Brainwashing*

"The privilege of a lifetime is to become who you truly are." Carl Jung

I would add "and be valued for and surrounded by people who are interested in who you are."

A crisis sorts out what is important and who is important.

Consider this hypothesis:

We are ALL flawed and highly reliant on each other, and we can afford to isolate our "enemies."

From a Christian perspective, Jesus presented the toughest challenge for us, we should "Love our enemies".

It is the wrong people who are suffering.

People who care and are losing the courage to speak out, and when they do speak against the common wisdom they get attacked and unfairly labelled. This is leading to a moral decline because those that don't care, prefer to smear, destroy, and come to the fore, while the wrong types begin to thrive. This requires new skills to cope with and navigate too, captured in our book *The Evil of Silence.*

Getting in control of technology

Whilst the issue with technology is a psychological one, causing this mass psychosis, the notion that technology will self-correct and laws realign to meet our needs, is flawed because technology is unlike humans, is tireless, moves too fast, is irrational, and impacting at a time when we have never really needed to understand humanity, let alone the psychological impact of technology on humanity.

And or unlike Hitler or any other tyrant, they eventually got found out and usurped by the masses propagating their demise.

In the case of technology, a demise of technology is unlikely because in a tech-fuelled society our ability to organise ourselves and have meaningful debate across divides is rapidly diminishing, conflict is increasing and so is free speech.

Such binary thinking, polarisation and echo chambers are akin to people imitating technology and becoming more like machines?

So, the challenge is raised in terms of what is the source of morality and what is right or wrong.

This must go beyond making money on the crazed valuations of bitcoin, and secular leadership strategies.

This parasitic type of tech investment is taking cash, investment, and resources from and weakening the productive economy and defi/crypto for example reducing solvency of the banks that caused the 2008 banking crash too.

The hierarchy of business ethics is driven by:

- Truth

- The society impact

- Faith/ethics

The attempts to define a personal secular sense of "good" morality leads to idolatry and conflict,

and those that fall outside these ideals are "evil." This is a dangerous, highly sensitive territory because ethics based upon faith or any other woke criteria are subjective and difficult to rationalise.

In 2012, likes and shares changed everything on social media, and its social impact has become far more harmful.

So social media fuels this activism, echo chambers, and moralism, leading to blind blindness where truth is not valued nor is it emotive, motivating, or appealing.

"There is no coming to consciousness without pain." Carl Jung

The pain and or difficulty in doing the RIGHT thing, can be confused with doing the wrong thing so fall foul of and be confused with the woke notion of taking offence, so one can be falsely accused of denying the “lived experience” unable to action a sensible solution!

This raises the question what are we as human beings?

We are love machines, machines; the tricky thing is this:

"What do we love?"

At the very top level, we have a choice to follow the first commandment “to love God and other people”. Jesus said there is no greater law than this.

There are dramatic implications and challenges that Jesus went on to identify:

• Love thy enemies

• Forgive and show grace

The most realistic place for us to experience “imposter syndrome” or feel constantly humbled.

What good looks like paradoxically is to realise that we are all flawed and we do not know better.

The most realistic, safest place for us is imposter syndrome or constantly humbled or to realise that we are all flawed and do not know better. Humility is the most natural place to be because we know so little of what there is to know, and far more that we don't know. Open your eyes and see the beauty, look outside the window and its impossibility around the detail: why or how do leaves decompose?

Socrates evidenced that our survival and success are entirely dependent upon others. Others to challenge our thinking, and or wellbeing based upon our willingness to listen, take an interest in different views and support each other.

Grace and love are critical to avoid conflict and the means to quell anger.

So, who do we turn to, a tolerant model or a secular idolatry-driven model that leads to idolatry?

Idolatry happens if we look too low and or ignore the fact there is an infinite source of higher wisdom or a God. Conscience or wisdom is clearly external and ethics a heart-driven thing that doesn't, as Google noted with their Oxygen Program, fit neatly with typical metrics.

We tend to create idolatry even though it may sound great, therefore it is incomplete so has unintended consequences. The negative unintended consequences, which we can see are all around us affecting:

• Mental health

• Social media and driven conflict

• Sustainability and consumption

• Tech-driven satisfaction like / share usurping our natural face-to-face systems

• Free speech

• Fractious relationships

• Profit

• Tyranny control

• Critical thinking and healthy open debate and the value of truth

• Woke mantras victimhood etc

Social systems are checked by teams, whereas tech is vulnerable to waves of disinformation below:

So, who can or should set morals – God or us? If we think we know better and set them, history suggests this tends to lead to war, and utopia tends to destroy foes.

What can we rally around?

We seem to have a God-shaped hole and if we were made to be addicted, were we made to be addicted to Jesus, not or digitalisation/social media. Addiction to social goes some way to explain why we may be coming and or become more akin to technology than people?

Dr. Mate believes that addiction (or what we love) is not a choice, neither is it all about drugs and illicit substances, addiction affects most of us – whether it be to alcohol, nicotine, sugar, work, or exercise… the list is endless. He calls for a compassionate approach toward addiction, whether in ourselves or others. As he puts it, the question we need to ask is not why the addiction, but why the pain.

Why do we now live in a culture that doesn't meet our human needs?

It seems that technology is an addiction: *"addiction is manifested in temp pleasure that therefore we crave with negative outcomes in the long term and inability to give it up."* Dr Gabor Mate

Yet, if the outcome of loving God is good for us and we are addicted that's surely a good thing?

And if the idolatry or mantras are ones of intolerance, that's sure limiting personal development, truth, beauty, and or fuelling conflict?

Liberal democracy is a tyranny of tolerance. Because forcing people to be tolerant is forcing people to stay silent and go along with subjects that have a multitude of different views: such as abortion, gay marriage, or the multitude of identity politics-related perspectives.

If we can rally around one tolerant God that is alive, knows better, and had sorted this, life would surely still be challenging but far easier? So, if we loved that God and other people, we probably wouldn't go too far wrong!

Rational factual approaches to trying to define secular society moral codes are just boring.

So it seems that our God, such as Jesus, must be someone or a person so that we can emulate and imitate in order to be interesting or engaging?

Inanimate truths are of less or no interest, unengaging, meaningless, and just boring or in the case of China boring and tyrannical?

President Xi forces this ideal on Chinese Christians and Uyghurs by forcing them to memorise his speeches, almost putting Xi himself in the position of a God-like role model!

Role models

We need to have something to worship, and it seems that it must be a person or human role model to engage us.

Technology, however, seeks to control our attention and so manipulate us with devastating implications on our wellbeing and personal journey of self-discovery.

It's not easy either, and or we need a mentor which in the Christian tradition is Jesus. We look for role models in life although Christians are flawed role models because they naturally fall far short of Jesus who was God.

"The sad truth is that man's real-life consists of a complex of inexorable opposites—day and night, birth and death, happiness, and misery, good and evil. We are not even sure that one will prevail against the other, that good will overcome evil, or joy defeat pain. Life is a battleground. It always has been and always will be; and if it were not so, existence would come to an end." Carl Jung

Bad ideas dominate engagement on social media because we don't react or respond to positive news, so bad ideas tend to dominate online ethics and these tech online echo chambers.

This (negative) feedback loop leads to depression, no thinking, and "just doing." Doing is subconscious- driven activity, that involves no thinking, so we lose sight of a common good.

Gen Z has been impacted most by the "like and share upgrade" to social media which has been so harmful and caused their mental health to suffer the worst. Their thinking is dominated by trivia rather than learning about anything useful beyond their direct social influence, so they are trapped.

This trap of negative-positive feedback, conflict, the dominance of negativity, reduced thinking, the threat of being labelled as the evil one, and the trap is a threat to social democracy.

Yet materially, things are getting better since the 1970s. In the 1970s there were sensible material reasons for crises such as the 1973 Oil Crisis. Now the reasons for the crisis are less obvious, and we are losing sight of why we are unsatisfied or what "good looks like."

It begs the question of whether the post-2012 generation will ever experience what "good" looks like, and if so, how might they experience this?

Valuing truth should be the tool to irradicate falsehoods, but people find meaning and purpose in these falsehoods like postmodernism; the BLM movement; identity politics; and wokeness or victim culture that tries to conveniently destroy truth on the basis that science is just another of man's fiction.

These are general (and exclusive so divisor) mantras such as "Black Lives Matter" and or "stable coins." Generalisations are attractive and create an almost God-like position because they create certainty, fervour, and purpose fuelling dismantling of faith-centric certainty and the attractiveness and growth in tech-enabling echo chambers and of a self-righteous high ground during the pandemic.

Yet God is the probably only constant to provide certainty.

People cannot operate without certainty. Indeed, people will knowingly take the wrong action to reach certainty because they know it is wrong and unchallenged. Although when challenged, as the action is wrong, it's undefendable, and they smear critics and get angry to defend their purpose, position, and passion.

Reaching the truth has become filled with fear in this terrifying social media environment because people who challenge China Wuhan lab as to the source of the virus, have been accused of being racist. Racism is a vicious accusation, and of course flawed, and the Wuhan source is a sensible challenge.

Wikipedia has shown how to create the best online content: created from different viewpoints, and non-profit is the best way to ascertain a sensible, dispassionate, and truthful outlook.

Bad news spreads faster than good, and when the statement has no context such as the 140-character tweets on Twitter, it's a problem because it's neither defendable nor possible to make sense of something so inflammatory.

Atonement means “at-one-ment” which is the intimate connection between people for example in a jazz band, or soldiers fighting in the war, or the means to physically and mentally feel physically and mentally great.

Bonding together is an essential part of civilised society and religion; the key is who is welcome or unwelcome to these groups. Christian faith should welcome everyone with grace.

Grace is the ability to put the past behind, forgive and move forwards without holding a grudge.

Grace acknowledges the human plight, so is critical to maintain wellbeing and reduce anger.

It is when we feel loved and valued for who we are that we perform at our best.

The conditions for peace must be law-abiding and a shared simple top-level ideal, such as “love God and other people.” And then we need to ACT on it.

In this group, like Wikipedia contributors, people have a chance of exploring the truth through open truthful speech.

The alternative is imitating a lower level unengaging ideal, and if it's secular, not personified, we CANNOT imitate it, so it’s useless.

So, the business hierarchy is

- **Truth**

- Social factors

- Faith ethical factors

Untruthfulness is probably becoming attractive to people when businesses become or are meaningless, because people can find something in some emotive mantras: certainty and rally behind, regardless of whether it is right or wrong.

Paraphrasing Jordan Peterson,

"Almost all of our ideas are wrong, and unchallenged ideas become our truths."

So productive, open, challenging dialogue is critical to stay on track.

The neo-Marxist mantra with power being the overarching corrupting force in society is worth arguing with because of its lazy thinking and label. But it's worth noting that it’s a generalisation and negative generals are all flawed. So, if the debate ensues on the basis that it’s a useful starting point it will go nowhere because it’s a generalisation without context so will lead to conflict. So, ask, "can you give me some specific examples?". Then a sensible discussion can take place to avoid conflict of general vs generalisation.

So, what are the options – tech utopia?

Utopias such as defi are dangerous because they require strict conformity. The c19 violent history of Marxism shows the need for conformity and suggests this conformity leads to violence. Silencing disparate views leaves no other option other than violence to communicate.

So, in a defi society who is in control, or who do you want to be in control, techies, or a democracy?

Crypto facilitated by the blockchain structure is a subset and the building blocks of a tech utopia. The mass psychosis of the echo chambers, are fuelled by underlying mantras with a notion such as that "corrupt governments have been deflating your currency and wealth for years."

Yet there are, at times of recession, when printing money makes sense to keep the economy from collapsing and causing a far larger crisis and or mitigating unnecessary damage done by a downturn, that with funding, time will resolve. Yet whilst the question is ambiguous, 92% of the respondents to this survey wanted a decentralised solution.

Historically, governments have at national level, controlled currency and distribution of wealth via taxation, and interest rates, and to a degree managed debt and currency valuation via inflation.

In democracies, this has been and should be targeted with the broader interests of the population and the vulnerable in society in mind. Democracy allows for the fact that if the Government messes up, they get voted out.

The notion of defi self-governance, and that you will remain in control in a decentralised finance defi-centric society, is nonsense because there will have to be governance anyway.

With a centralised defi society the governance structure is far more tyrannical with atomised "with control" at the individual level. In order to ascertain voting rights on the myriad of subjects' governments look after in a defi society, people will have to be measured upon some sort of social scoring system defining who will be allowed or qualified to vote or not.

This tyrannical misalignment of actions and beliefs is the cause of stress in people, and once normalised in a corporate or culture causes mental illness and psychosis.

Attacks on identity

Tech-fuelled fragility and the loss of identity caused by identity politics propaganda undermines the basic facts of life:

- the beauty of the male and female role in society

- mothers and fathers

Identity politics is ultimately not about diversity. The outcome of challenging the basics of your identity means a complete loss of your purpose and leaves you nothing to fight for. This leaves people with integrity who care and run into the cancel culture, vulnerable to suicide. People who lose themselves become fragile; how will our Gen Z kids ever know themselves?

This menticide has crept upon us. Waves of terror and isolation have made us aware of the importance of relationships, but notice how the same woke and tech tyrannies still prevail in the residual silence?

This leaves so many CEOs with their actions and beliefs misaligned unaware that their flawed panicky tech strategies and loss of meaning are costing their people a mental health crisis.

Investment in technology for the first time in 40 years has become the biggest investment category having just overtaken oil. So, it's no mean challenge and has a massive PR / menticide machine behind it!

Who is accountable for defi? Can you vote against the defi decentralised option? If crypto crashed right now there is NO recourse for "investors."

The cost of speaking out against these technology-fuelled mass psychoses is alienation. This alienation is without recourse to the law or accountability, and in fact, usurps the law. The impact is that your phone stops ringing, you may be labelled as being toxic, ostracised, and your online past "informally investigated", thereby usurping the law.

Lily Hajdu-Gimes, a Hungarian psychoanalyst, diagnosed the trauma of forced conformity in patients, as well as in herself. "I play the game that is offered by the regime," she told friends. "Though as soon as you accept that rule you are in a trap."

What's the leadership / societal challenge?

We are our worst enemies, and our greatest threat is when mental illness becomes the norm, or a psychic epidemic or mass psychosis takes hold.

The rise of the victim leads to tyranny and menticide and a hopelessly vulnerable state.

Consider "are you in a hopelessly vulnerable state or too vulnerable or fearful to challenge the woke mantras?"

Mass hysteria or mass psychosis is an extreme form of groupthink when members of the group follow the consensus instead of thinking critically and valuing the truth and healthy open debate to inform decisions.

George Orwell said, *"in a time of universal deceit, telling the truth is a revolutionary act!"*

"What is needed now is a swift cold shower of reality. The time for self-harming distractions with "wokus pocus," obsessing about sexual dysmorphia, Marxist "critical race theory" and "climate catastrophism" ... - all of which the Chinese Communist Party is glad to encourage - must end." Gatestone Institute 19/8/21

What are you too afraid to discuss that you know needs to be challenged and thought through?

How do false truths and ideology gain traction?

Illusion or ideology gather support during tough times of non-reality because ideology affords people the luxury to deny, blame and distance themselves from reality. Victimhood is a convenient position. The falsehood is convenient and provides a means of creating certainty and unity with others, and a radical spirit, drive, and purpose.

Falsehoods often dominate ideology and gather traction because they normalise and provide flawed: purpose; group identity; and a weird sense of moral high ground.

People can't cope with uncertainty so no matter how weird the ideology, it creates a sense of certainty and moral virtuousness and righteousness, even when the bases of the ideology are patently false.

The falsehood creates a blind spot, which closes the counterargument.

Certainty is particularly attractive when all else is falling apart. However, the flawed ideology typified, highlighted by the predominant mantra or generalisation, is always at the root cause of a crisis.

This is psychologically the cause of a company turnaround, where the generalisation creates a blind spot. This distortion or deletion causes strategy to miss the real issues, by distorting or deleting the facts. Most "company turnaround practitioners" and business leaders completely miss this mechanism so get duped by the comfortable mantras and lazy thinking that beautifully justifies the status quo.

People will always do all they can to do what they want to do and be left alone.

The misleading language

Marxism used to be focussed on the economic oppression of the working classes. This focus lost its relevance so now the neo-Marxist postmodern tools and mantras have been replaced and centred around emotionally charged subjects such as race, sexuality, or minorities. This is where people can choose to feel marginalised or hard done by or victimised. This victimhood leads to scapegoating around very divisive clever general mantras, lacking specifics, and context. This divides groups.

Woke language is clever at fuelling division, a moral high ground, and exclusivity such as Black Lives Matter. The next level of distortion critical race theory when looking at everything through a racial lens, with white people being the perpetrators so unable to comment, is patently nonsense untrue, as well as rightly angering good well-meaning white people with the appalling false accusation of implicitly being racist.

Specifically, do you see how the headings “Black Lives Matter” and “critical race theory” or “postmodernism” create division and distortion?

Penalties

The penalties that maintain the tyranny and silence debate

The scapegoats that infringe the ideology get an aggressive, emotive, personal protective dialogue and with vicious slurs and labels. These vicious labels are designed to scare off challengers.

Specifically: how do you respond to these vicious flawed labels when you challenge these mantras? What EQ skills do you need to do that and what is the awareness to develop our skills?

Yet our hubris shows we are unaware of the need to develop our strategies:

Do you see how the underlying thinking in the dialogue and these terms are used to attack and create exclusive echo chambers and attack conflicting views from the moral high ground such as Christianity or any other faith?

How do you challenge the ideology?

Our natural inclination is to reject ideas that don't fit our taste and defy error that damages our illusion, mantra, or ideology.

These mantras can be seemingly positive such as rejecting views that don't fit with your version of the truth. This creates blind spots.

Whoever challenges woke illusions conveniently becomes a victim of scapegoating, and so will be met with anger. Anger because you are challenging the fundamental woke or another anchor, which could be that person's means of "coping or finding meaning, woke group religion, and or of course the pain of denying they could be wrong."

Specifically: Do you see truth as a journey and take an interest in finding out the merits of other people's conflicting views, before taking your view?

Why is a challenge to ideology met with anger? In their heart of hearts, they know what they are doing is wrong so can't sufficiently articulate a response to the challenge, so anger is a convenient and last form of defence.

The ideology trap is when we lose touch with reality and live in a delusional state. So we habitually look for information that fits with our delusion and therefore reject anything else, ***without thinking,*** leaving us vulnerable to:

• social media echo chambers

• negative emotions

• panic

• isolation

• fear

Conforming actions creates waves of fear so maintains tyranny.

You might not see how using “pronouns” could fuel evil.

I added pronouns to my header because “I could not see a downside”, or I’m “not sure why I should *not* use pronouns.”

Acting by adding pronouns forces us to realign our beliefs and get drawn in. This is akin to saluting Hitler or what is underlying the ideology. This draws us in, and flawed actions trap us.

It starts with something baseless and flawed and part of woke ideology such as the pronoun aspect of identity politics. The iterative tools of tyranny are:

• to create waves of fear

• nonsensical waves of propaganda that make no sense

• digitally deselecting information

There is a link between adding pronouns and or saluting Hitler. Both are wrong and falling victim to tyranny, whilst the methods were different, the victim, scapegoating and terror/silencing processes are the same.

The woke terror now is found on social media, in the form of marginalisation, public attacks, and labelling. Marginalisation on social media creates the equivalent emotional and physiological response akin to being physically attacked.

Tech ideology and mantras such as “tech is the future”, “digital disruption”, “governments can’t be trusted” have led to the panic and corruption of tech investing, which are often drifting no more than to become a Ponzi scheme.

Most of these Techcos lack meaning let alone a compelling proposition, yet the product, system, social media, and information are subject to manipulation using diversionary general self-fulfilling language, such as “decentralised finance” or “*social* media”!

The perceived valuations of these often-loss-making tech entities are driven by burning cash to inflate valuations, and a focus on growth, and domination.

During this process, professionals and leaders get duped so become compromised, so morality becomes compromised, and greed takes hold.

Many are unaware of the tech menticide or collective madness that is occurring, leading to confusion, and panic. But it is hard to admit and reverse.

Trapped in the present.

People are trapped in the present because of their addiction to social media. They are unable to step out of the endless propaganda leading to some weird outcomes and waves of untruths:

- postmodernism defines science as a man-made construct
- men can get pregnant
- gender pronouns
- lived experience
- taking offence
- critical race theory
- tech is the future
- victimhood is in control
- labelling: racist, sexist, transphobic

Propaganda weakens people and their resolve.

Propaganda creates psychosis because it's delusional. It weakens people and makes people vulnerable to a saviour such as Xi or Jesus.

"It is not for nothing that our age calls for the redeemer personality, for the one who can emancipate himself from the inescapable grip of the collective and save at least his own soul, who lights a beacon of hope for others, proclaiming that there is at least one man who has succeeded in extricating himself from that fatal identity with the group psyche. For the group, because of its unconsciousness, has no freedom of choice, and so psychic activity runs on in it like an uncontrolled law of nature. There is thus set going a chain reaction that comes to a stop only in catastrophe." Carl Jung

This propaganda-fuelled psychosis eventually creates menticide which is when totalitarianism takes hold. Currently, menticide is manifesting in many forms.

- woke ideology: postmodernism

- China Communist party ideals

- Defi utopia, Bigtech ideology, cryptocurrency

Totalitarianism is the greatest threat people face. It is based on a flawed assumption that the ruler is themselves, another human, and NOT God. This does not fit with the flawed nature of people; people are flawed and know so little and are therefore completely incapable of fulfilling this role.

Total power corrupts, and the masses become childlike, transforming sound minds to sick minds. As a check, "what is the alternative view to this propaganda"?

• the use of pronouns

• gender usurps biological sex

• "tech is the future"

• crypto: "FIAT currencies and government institutions are corrupt"

The last point is one of a culture turning upon itself, causing good-thinking, well-intentioned people to turn on and blame their governments and institutions, rather than take personal responsibility.

The issue is the same: victimhood. Blame and propaganda signify and maintain victimhood.

Ideology creates a childlike state.

Totalitarianism leads to mass suffering and social ruin. Ideology, rather than free thought, dominates and becomes accepted, killing the mind or menticide.

This self-righteousness with a set of attack tools is a comfortable place for people or groups who feel victimised, proving the opportunity to cling on to flawed ideals, such as wokeness. From there, there is high ground and battles to fight from which they attack others, rather than take responsibility for searching out the facts.

This is a childlike state operating in tyrannical ideals. Self-righteousness is found by attacking those that don't fit with these ideals.

Disgust and waves of terror and the stupid mind

Hitler often used words of disgust when referring to the Jews. Disgust makes it easy to isolate responsibility and obsessively scapegoat and wipe out your targets. Manmade, moral matrixes like these are dangerous and nonfactual territory that create conflict that's nearly impossible to rationalise.

It is more comfortable to believe ideals than to face the facts. Facing the facts requires skill, intelligence, courage, and competence to action.

Waves of terror interspersed with periods of calm, cause morals to sink and soften people up to become manipulatable and non-thinking.

Propaganda is comfortable because it provides meaning and normalises this madness and creates a weird sort of meaning and purpose causing once good people to become revolutionaries for the ideology.

Like postmodernism, that believes truth is a man-made construct, and the second wave lie "lived experience usurps the truth."

Lived experience is tough to challenge because it's illogical. Although a half-truth without context is so more appealing than reason. It gives the cowardly, stupid, and less able mind real power, gained through a self-righteous ideology and a set of tools, and ready-made team of supporters to attack and cancel those who dare to challenge what doesn't fit their ideology or can see through the woke lies.

Turning on your own and isolation.

The addictive nature of social media and fear of attack causes people to become vulnerable to propaganda and turn on their own. This is how to obfuscate and hide the truth. For example, CCP members are openly attacking America for their errors now on LinkedIn, under a higher moral guise Xi! The ultimate totalitarian. LinkedIn is the only social media approved by China.

"Indeed, it is becoming ever more obvious," he writes, "that it is not famine, not earthquakes, not microbes, not cancer but the man himself who is man's greatest danger to man, for the simple reason that there is no adequate protection against psychic epidemics, which are infinitely more devastating than the worst of natural catastrophes." Carl Jung, *The Symbolic Life.*

". . .the totalitarian systems of the 20th century represent a kind of collective psychosis. Whether gradually or suddenly, reason and common human decency are no longer possible in such a system: there is only a pervasive atmosphere of terror, and a projection of "the enemy," imagined to be "in our midst." Thus, society turns on itself, urged on by the ruling authorities." Joost Merloo *The Rape of the Mind.*

When we can't cope, we will normalise things to make sense, justify inaction, and cope in a vain attempt to justify the status quo.

Very few saw the rise of Hitler as being a problem, yet a few did.

Language is deceptive and changing it is powerful: rather than connecting us, social media isolates us. Isolation is the perfect tool to disrupt the healthy dialogue interactions that keep us in check. These healthy interactions enable us to see through propaganda. Isolation, however, enables new patterns of thought to be formed and tames wild animals.

Isolation in combination with waves of terror or uncertainty, cause many people to reach a hopelessly vulnerable state. In this state we crave certainty, which can be any form of certainty, many will make bad decisions to reach a point of certainty.

Action is very visible and always imperfect, so tough times challenge morality and social norms.

This causes good people to attack the institutions that keep them safe. So, ideologies like “FIAT currencies have robbed the masses” or claims that “governments are corrupt” justify cryptocurrency and create blind spots preventing with no thought as to the costs, broader social, or moral considerations.

This ideology is convenient for investors who are panicked by having nothing good to invest in, by pumping up the valuations of cryptocurrency. Non-compelling untrusted fintech models are ideal.

These tech businesses largely give away their services with losses fuelled by ever-increasing equity investment rounds that bump up valuations. These are very like a Ponzi scheme version with each investment round passing the parcel. These valuations are based upon and fuelled by the internal sub ideologies of these tech businesses, creating their own unchallenged echo systems and convenient terms “bitboys” or “nocoiners” to describe their respective echo chambers to fuel their ideology and isolate others.

So, given the priority is to stabilise the economy post-pandemic, cryptocurrency is the antithesis of a solution, ill-thought-out, raising more questions about its fundamental implications that require a fundamental review versus the unspecified gains through streamlining bank processes using blockchain.

• Liquidity is a big issue with crypto in relation to drawing cash away from fractional reserve banking mechanisms in the retail banking sector.

• The ever-increasing power consumption of bitcoin alone increases and is currently 50% of the UK’s power consumption. Ever-increasing because this is related to the increasing escalating and unstoppable cyber-security concern risks from China?

• There is an onslaught of lies and assumptions about crypto such as "stable coins" will be underpinned by assets, drawing away further liquidity from the productive economy?

• Crypto "stable coins" are surely another misleading mantra and a contradiction in terms?

• Volatile valuations will drag investment away from FIAT currencies undermining solvency -

the opposite of the mitigating bank balance sheets post 2008.

• The increased reliance on digital will leave the UK wide open to cyber-attacks from China.

• Paraphrasing the chair of the UK APPG on Ai discussing crypto adoption said last night (27

Oct 2021), he quelled a discussion following a few tough questions from doubters raising issues re crypto viability "well it's happening".

"he who dictates and formulates the words and phrases we use, he who is master of the press and radio, is master of the mind."—Joost A.M. Meerloo, *The Rape of the Mind: The Psychology of Thought Control, Menticide, and Brainwashing. "Where thinking is isolated without free exchange with other minds and can no longer expand, delusion may follow. Whenever ideas are compartmentalized, behind and between curtains, the process of continual alert confrontation of facts and reality is hampered. The system freezes, becomes rigid, and dies of delusion."*—Joost A.M.Meerloo, *The Rape of the Mind: The Psychology of Thought Control, Menticide, and Brainwashing.*

"There is another important weapon the totalitarians use in their campaign to frighten the world into submission. This is the weapon of psychological shock. Hitler kept his enemies in a state of constant confusion and diplomatic upheaval. They never knew what this unpredictable madman was going to do next. Hitler was never logical because he knew that that was what he was expected to be. Logic can be met with logic, while illogic cannot—it confuses those who think straight. The Big Lie and monotonously repeated nonsense have more

emotional appeal in a cold war than logic and reason. While the enemy is still searching for a reasonable counter-argument to the first lie, the totalitarians can assault him with another."—Joost A.M.Meerloo, *The Rape of the Mind: The Psychology of Thought Control, Menticide, and Brainwashing*

Ideology provides certainty.

Tech, Woke, and China versions of totalitarianism tragically provide this certainty when people are at their wit's end. Divisive language creates a revolutionary zeal caused by a good society to turn on itself when things are too easy. This strategy is far easier than taking personal responsibility.

It's far easier and more comfortable to be a victim than to face your failings and find the courage to consider and face the future and plan accordingly.

Jesus said, "The proud are an abomination" and for this very reason they have no chance of turning to their saviour.

"I am the way, the truth, and the life. No one comes to the Father except through me" Jesus said in John 14:6.

The scale of the challenge is daunting because at its most basic level it's a battle between:

1. Growth and domination – the current insatiable madness

2. Profit and relationships – back – basics - realigned mantras

Bitcoin and Silicon Valley have a whole anti-establishment aspect to it, and this creates an odd link with symptoms of the CCP tyranny that are trying to undermine Western values and institutions.

However, the secular outcome is a pathological order with no spontaneity, curiosity, no beauty or joy.

So, what could prevent this menticide?

Create a parallel structure to:

- reset relationships
- rationalise businesses viability and valuation based upon profit
- enable the right people with integrity to thrive
- face the truth – tools to identify and understand the distortions
- challenge ideology to check truth and validity
- open dialogue and identify the blind spots
- create personal responsibility
- create empathy to acknowledge others
- value different views
- take lots of self-realised actions
- provide grace, forgiveness, and love

Wimps and cowards thrive in a woke culture because they don't have to face reality or take responsibility for their actions so can only focus on avoiding debate, attacking, and destroying.

Woke language is clever at fuelling division, by creating a moral high ground and exclusivity victimhood such as Black Lives Matter. The next level of distortion is critical race theory, which looks at everything through a racial lens, with white people being the perpetrators unable to comment.

This patently untrue nonsense also angers good, well-meaning white people with the appalling false accusation of implicitly being white and therefore being a racist!

Specifically, do you see how the headings “Black Lives Matter” and “critical race theory” or “postmodernism” create division and distortion?

The neo-Marxist mantra that power is the overarching corrupting force in society is worth arguing with because it's lazy thinking. But it’s a generalisation, and negative generals are all flawed. So, if the debate ensues on the basis that the pretext is a useful starting point, it will go nowhere because it’s a generalisation without context, which will lead to conflict.

“Abuse of Power?” The solution is to ask, “can you give me some specific examples?” Only then can a sensible discussion take place as a rational means to avoid conflict of generalisation vs generalisation.

How do you challenge the ideology?

Our natural inclination is to reject ideas or perceived threats that don't fit our taste and defy error that damages our illusion, mantra, or ideology.

These mantras can be seemingly positive such as we are the "Best in Class." They create blind spots.

Whoever challenges woke illusions conveniently becomes a victim of scapegoating, so will be met with anger. Anger because you are challenging the fundamental anchor, which could be that person's means of "coping or finding meaning, or the pain of denying they could be wrong."

Specifically: Do you see truth as a journey and take an interest in finding out the merits of other people's conflicting views before taking your view?

Why is a challenge to ideology met with anger? In their hearts of hearts, they know what they are doing is wrong but can't sufficiently articulate a response to the challenge. So anger is a convenient and last form of defence.

By taking lots of small actions, you break the tyrannical cycle. This enables people to realise the value of their peers and the value of themselves to their peers, so, the outcome is self-fulfilling.

For example, by putting others first you create a self-virtuous cycle but only if self-actioned rather than purely taught. Experiential learning is enabled by taking new action from a different perspective.

This is the way we learn. See, you have the infinite capability. The infinite element is through a connection with God or else your limit is you!

The best place to be is somewhere like realising that we are all flawed, and no human has the ability to act as God and set the rules, judge morality and or set the way anyone thinks.

Businesses in turnaround scenarios descend into tyranny and exhibit the same traits, and the appropriate response is not more tyranny. Rather, people need love, oneness, reconnection, and clarity, not more dictum to resolve what has become unconsciously for them, their self-made psychology, that has created their own impossible scenarios.

Correctly facilitated with the correct perspectives this new self-realised action immediately gets results, because immediate new actions are taken, courage is overcome and so perspectives permanently shifted.

It’s a difficult process, but once the ball starts rolling it's exciting like a new flower coming out or a jazz band that can fluently play a completely new set of improvised tunes. TOGETHER.

By using winningthinking®, we have got businesses back to profit and cash in 2 weeks and cash weeks from haemorrhaging £400k/month. So new action works immediately, and the new shifts create immediate results, cash, profits, and refreshed relationships, and reset mental health. It fixes the right problem that drives everything: the perspectives and the thinking.

This realignment restores and creates new energy.

But look at the action CEOs are taking, fuelled by the Big 4, leading to more panicky fractious investment in technology. Technology is the antithesis and fuels menticide and mental ill-health.

The risks are high because people are trapped in a tyranny faced with a personal attack for standing against these misnomers. If people are attacked it is tempting for them to apologise and fall in line. This causes people to compromise and lose themselves and their souls or their strength to cope.

This tyrannical misalignment of actions and beliefs is the cause of stress in people, and once normalised in a corporate or culture causes mental illness and psychosis.

Tech fuelled fragility and the loss of identity caused by identity politics propaganda undermines the basic facts of life:

• the beauty of the male and female role in society

• mothers and fathers

This is ultimately not about diversity the outcome of challenging the basics of identity is a complete loss of purpose and nothing to fight for. This leaves people with integrity facing the cancel culture vulnerable to suicide.

New action

New action from re-aligned paradigms realigns actions get back to humour, new action, freedom, and intimacy.

The drip-drip effect.

When to act, it's TOTALLY down to you!?

NOW is the time to take new action. Do you have the skills to face this and realign action?

The Harder the challenge the greater the glory.

It's down to YOU…

Photo Source Unknown. *Anonymous source thought to be a rally in a German dockyard in 1935.*

August Landmesser was the German that had the nous not to salute at the German rally in Germany in 1936 when a newly built ship was about to leave Hamburg harbour. The photo shows a crowd of people, who are believed to be workers, raising their arms for the infamous saluting gesture, except for a man believed to be August Landmesser who is standing with his arms crossed.

The bizarre paradox is that whilst the pandemic initially caused us to realise the importance of people to– people relationships during the lockdown, acknowledging and making time for people, this is fading. I expect this inactivity and misalignment could cause the next wave to be more damaging to our well-being.

The impact of this menticide is far-reaching, but who is really thinking ahead or planning?

The pandemic has also provided a crisis that we have not addressed its root cause and potential source and blatant cover-up in China.

Take EVs… "When talking about saving the planet by replacing internal combustion vehicles with EVs, we first need to comprehend the scale of this initiative.

Talking strictly about vehicles, we need to face a simple fact, there are an estimated 1 billion cars on the road each year, and this number is increasing.

Considering that most of these vehicles are still using internal combustion engines, replacing all fuel-powered cars with electrics requires a lot of effort.

The most-produced car in the world is the modern Toyota Corolla, with roughly 43 million of them built over the course of 55 years. We do not have 55 years left to stop global warming.

Unless we take action to prevent the sale of used petrol-powered cars to the developing world, our best efforts of replacing traditional transport with the green alternative will soon prove to be in vain. Otherwise, individuals will be tempted to sell their old cars to foreign buyers, as governmental programs may struggle to make a more profitable offer.

While Europe will replace their current vehicles with electric alternatives, the developing economies in Asia and Africa are expected to proceed in the opposite direction.

However, comfort EVs do not come cheap, and most of the world cannot afford to buy a brand-new car without special government-funded programs. Not only are such governmental initiatives still rare, but most people would also be required to sell their current vehicle to fund the purchase of a new one.

The question is, what are we going to do with all the used petrol-powered cars?

Even so, car companies still need time to design, develop and market their upcoming electric models, similarly to how the legal infrastructure for the benefits of driving an EV required extended debate.

EVs are not bad. Maybe they are too late.

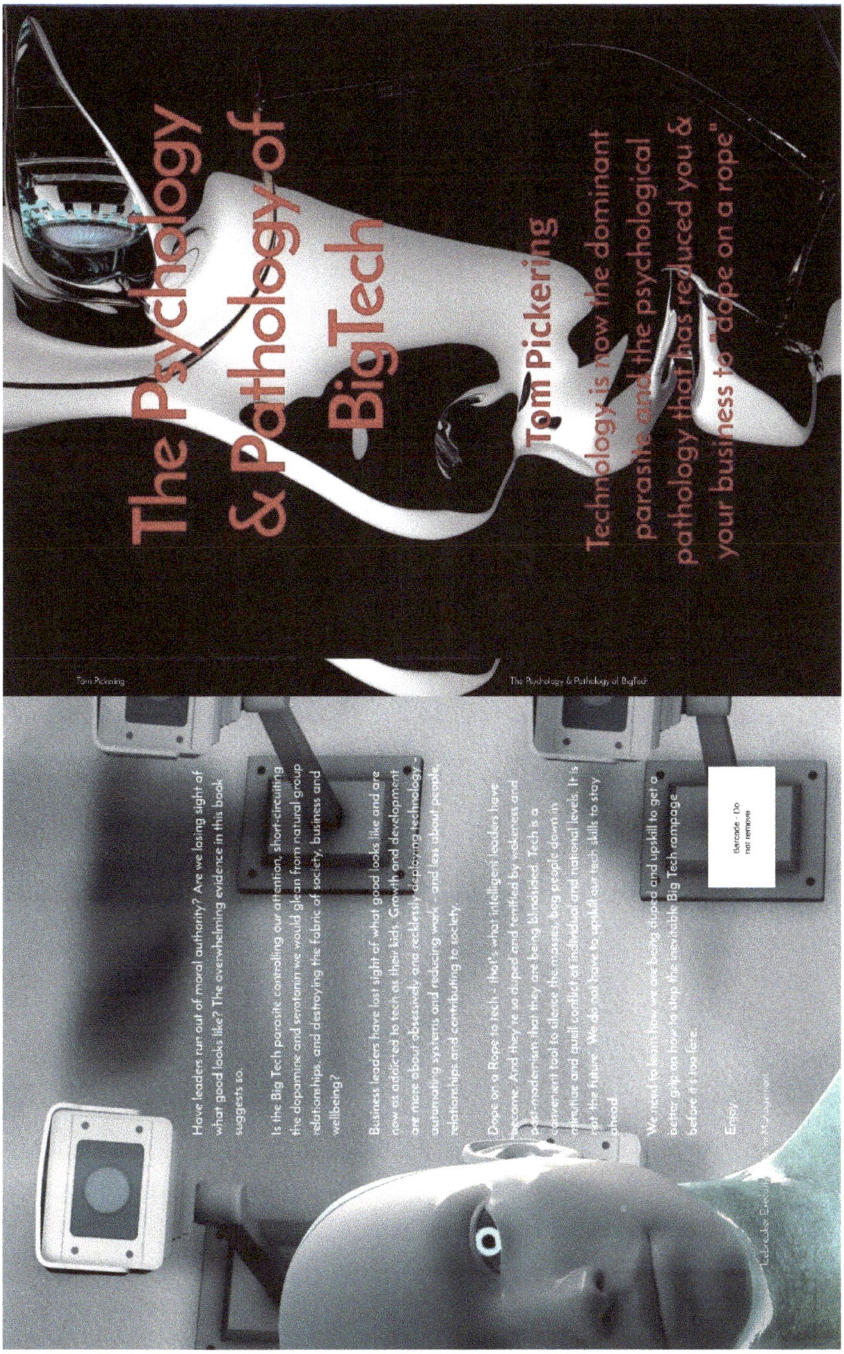
The Psychology
& Pathology of
BigTech
Tom Pickering
Technology is now the dominant
parasite and the psychological
pathology that has reduced you &
your business to "dope on a rope"
Tom Pickering
The Psychology & Pathology of BigTech
Have leaders run out of moral authority? Are we losing sight of what good looks like? The overwhelming evidence in this book suggests so.
Is the Big Tech parasite controlling our attention, short-circuiting the dopamine and serotonin we would glean from natural group relationships, and destroying the fabric of society, business and wellbeing?
Business leaders have lost sight of what good looks like and are now as addicted to tech as their kids. Growth and development are more about obsessively and recklessly deploying technology - automating systems and reducing work - and less about people, relationships and contributing to society.
Dope on a Rope to tech - that's what intelligent leaders have become. And they're so duped and terrified by wokeness and post-modernism that they are being blindsided. Tech is a convenient tool to silence the masses, bog people down in minutiae and quell conflict at individual and national levels. It is not 'the future'. We do not have to upskill our tech skills to stay ahead.
We need to learn how we are being duped and upskill to get a better grip on how to stop the inevitable Big Tech rampage before it's too late.
Enjoy.
Barcode - Do not remove

Technology is now the dominant parasite and the psychological pathology that has reduced you & your business to "dope on a rope."

Introduction

The Psychology & Pathology of BigTech: Technology is now the dominant parasite and the psychological pathology that has reduced you & your business to dope on a rope.

Have leaders run out of moral authority?

Are we losing sight of what good looks like?

Is the Big Tech parasite controlling our attention, short-circuiting the dopamine and serotonin?

We would glean from natural group relationships, and destroying the fabric of society, business, and wellbeing?

The overwhelming evidence in this book suggests so.

Business leaders have lost sight of what good looks like and are now as addicted to tech as their kids. Growth and development are more about obsessively and recklessly deploying technology – automating systems and reducing work – and less about people, relationships and contributing to society.

Dope on a Rope to tech – that's what intelligent leaders have become. And they're so duped and terrified by wokeness and post-modernism that they are being blindsided.

Tech is a convenient tool to silence the masses, bog people down in minutiae and quell conflict at individual and national levels. It is not 'the future'. We do not have to upskill our tech skills to stay ahead.

We need to learn how we are being duped and upskill to get a better grip on how to stop the inevitable Big Tech rampage before it's too late.

Forward

We cannot move forwards without sacrifice, much-emerging technology must be sacrificed.

Reckless - You are dope on a rope.

Technology is the dominant parasite and the pathology destroying the fabric of society, business, and wellbeing.

Business leaders have lost sight of what good looks like and are now addicted to tech. So they are obsessively and recklessly deploying technology: automating, reducing work, and naively deploying technology regardless of and without sufficient knowledge of the people or broader societal implications.

Intelligent leaders are unaware of how they have become dope on the rope to tech - fearful of getting duped by wokeness and post-modernism. These are convenient tools to silence the masses.

Silence suits the tech tyranny to quell conflict at individual and national levels as it unfolds.

Because leaders are duped into thinking technology is now “the future” they don’t consider that they might need to upskill to see this or cope...

So, guess what...? We are getting nasty results and losing sight of what good looks like.

• Mental health crisis

• Panic – increasing speed

• Environmental destruction

• Tyranny and silencing

So, this book is drawn from hard experience and many thought-provoking keynote speeches I’ve given over the last few years. I explain how BigTech type technology is simply parasitic and preys on anything of value: you, your business, and our wellbeing, and why we don't see this, and why we can't stop.

It's time to get back to business basics reset the paradigms that are driving this madness away from growth and domination to profit and relationships or this crisis will just escalate. A common response is "not knowing" – that's not good enough if we are taking actions, that in our hearts, we know to be wrong?

It's very tough to overcome our pride when we might have made a significant misjudgement.

Does this mean we might need to face up to the fact that we have made a huge mistake which puts us at risk of looking like we might have been duped? Maintaining good judgement is a real issue for leaders.

Yet the mantras continue, take the boss of LinkedIn discussing the playbook of scaling fast called "blitz scaling":

• World gets "more and more connected" (needs challenging – actually are atomised as individuals)

• "Global competition" (China don't compete – they attack)

• "1st to market and scale else you are fired"

• "Be part of the future"

Tech is taking over the biggest investment category for the last 40 years oil.

Without good judgement, we look stupid. But what is more stupid is not to face it, carry on and not change?

I observe that the "ever-increasing technological speed" is just a symptom of the stupidity of us not facing our errors. This is not the inevitable outcome we should just accept. Sometimes it is important to slow down think and realign to do what is right?

This book shares the broad picture that explains that our recent technological past and future could largely be the wrong path to take. This requires courage to stand up and say so when most are too fearful of speaking out. Therefore, I have written this book.

Government advisory e.g., the UK is heavily influenced by Techcos too:

What matters now?

What matters more than ever is not what is right and wrong, that's a journey, but to highlight what many might not see and a different view, from which to create some healthy open dialogue.

This affords us the opportunity to re-orientate, steer, and enact our strategies in the RIGHT direction.

We cannot move forwards without sacrifice. Despite the obsession and frustration with institutions, technology is not the answer. Much technology should probably be that sacrifice.

I am not proposing that the views in this book are correct. I am just raising and analysing the issues to maybe make us more aware of the challenge, and highlight the skills required to stay ahead and stay there.

The evidence seems to suggest that technology is changing our psychology.

What might good look like?

In this tech malaise and environment of ever-increasing speed etc, are we losing sight of what good looks like?

A Jazz band is a metaphor that we use to describe what good looks like, it's an intimate group of people in a group, not individuals. Tech singles out individuals and splits us across competing channels because there is less value (for them) in us working intimately together in groups, and we are more vulnerable to manipulation individually:

observe and enjoy the intimacy", "overflowing and engaging the customer", "taking new action without fear". "With a love for each other" to "hold the team accountable for excellence". "A high-octane human environment" "for breakthrough results", "on the job learning" and "new product development and profit".

What is the issue?

Meaningfulness requires us to face up to the worst of our past as well as realise potential by taking small new actions every day to improve our future.

The Big Tech mission, however, is to short circuit this, control our attention and hold us in the present. This tech world short circuits our dopamine and serotonin systems that we would get from natural relationships we would have in groups. This takes our sight off the past and off the future.

When you combine this with the threat of suffering for speaking up about doing the right thing, which should be our primary driver, and face a cancel culture, this misalignment puts us at risk of losing ourselves ifwe ever have had the chance to know ourselves. Do young people know what good is outside a tech world or will they ever realise, or is this now the unique privilege of those whose formative years were well before 2005?

The downside of this tech bubble/trap only really becomes evident when things get tough, or the future looks tough. So tech affords us the luxury to revert to embedding ourselves in the tech to escape taking responsibility and therefore action to do the right thing.

Like alcohol, tech triggers dopamine, so many are addicted to an “in their present fix” and stepping out of that tech bubble becomes increasingly scary. This is exactly where Big Tech wants us to control our attention.

Losing our purpose

The downside of tech is that our resilience is dependent upon us taking new action aligned with a strong purpose to make a positive impact on the day-day world by facing these tough challenges.

A purpose requires us to look to the future, which tech does not allow us to do. This is our unique ability as humans over machines to see and imagine this future as reality.

Yet trapped in tech bubbles, the tech is getting to know us better than ourselves, to anticipate our wants before we know them ourselves! Or manipulate/trigger a response by presenting emotive information.

We respond most to negative information and do little in terms of responses to positive information, so tech tends to pick up and feed that, which triggers us negatively because tech can't pick up on our non-responses.

Unawareness of ourselves

We are tragically vulnerable and unaware of what drives us. We don't have the self-awareness of how to adjust. This unawareness alongside manipulating tech mantras and language leads to us being subconsciously duped by technology.

"5G is more than just a technology, it is a platform for innovation. But players across the ecosystem need to work together to realize its full potential." Ericsson August 2021

The big unbeatable Big Tech bet

The basic principle and threat of Big Tech are based upon Price's law: the successful become exponentially successful and the unsuccessful exponentially unsuccessful. So, if Big Tech is constantly refining this formula, well beyond our human capability, we are in trouble.

“Daddy they won’t let tech get cleverer than us.” My daughter Rosie

Tragically I think tech probably already knows us better than we know ourselves now.

Broader tech issues

The other aspects of tech I discuss in this book - such as blockchain, bitcoin and others – have similar dystopian outcomes, driven by anti-establishment idealism.

Isolated individuals tend to form and maintain very narrow views. Productive interaction with others enables us day-day to see the threats and adjust accordingly. So, if we have been more isolated during the shutdown it is clear for us all to see more conflicting entrenched views and potential conflict even within families who are failing to communicate effectively because they are relying on more tech.

If we don’t get some sensible debate around technology and get a grip of tech, we are sure in trouble.

We are also opening ourselves to the wrath and dystopia of China and other tyrannical regimes with very different views of what a “successful outcome” looks like and different plans.

Tech does not allow for different views; platforms are regularly shutting down at the macro level and through low-level admin reasonable debate without considering where someone is coming from jumping to labels such as hate, racism, left, right, anti-this etc, etc.

The bigger picture

There are two related books to enable us to get a grip on the new now and make the seismic shifts and action to achieve the top 7 objectives.

It is important to understand how these top 7 are usurped by post-modernism or neo-Marxism.

So, I wrote a book on making sense of the dark side and facing the pitfalls that corrupt, cripple freethinking and open debate. The essence and title is *The Evil of Silence.* ISBN 978-1-9196337-0-1

Silence vs. doing the right thing.

Silence is what happens when good people become fearful and fail to take responsibility and act upon what they know to be right. Maintaining silence is how any tyranny such as that of Hitler,

Mao, Lenin thrives and through eventual complicity like tech, complicity is allowed to take root.

Tech is a form of tyranny because there is no space to cope with different views without extraordinarily inflammatory outcomes emerging.

Our logos enable us to be able to trust what we know to be right, you can feel that, but it is vital we openly communicate with everyone around us to share ideas and challenge each other to stay on track and healthy. It's only through open dialogue that even we have a fraction of a chance of staying on track.

Connection to the logos is acting in a way that is aligned with your potential rather than ideology, allowing you to realise "truth, courage and articulation", "intimacy", "connectedness," and at "onement." Intimate connection with others is the platform to enable this.

In isolation, or atomised by technology, we are far too blinded and only see a very narrow view of reality. Winningthinking®'s *You Have Infinite Capability,* ISBN 978-1-9196337-2-5 shares the tools to highlight the pitfalls, open those possibilities, and practice taking the action we need to, to make our lives slightly better for ourselves, others, and our environment every day. It is from this NEW action that we learn!

This book does not intend to tell you what to do: we are terrible masters and worse slaves; people never do what they are told.

So, the intention is to raise the issues, confront our potential to reach a point, and touch on the means to keep the dialogue open and healthy. Healthy, open dialogue is the only way to enable us to realise something good.

The antithesis is receiving punishment or being accused of rape culture racism or some other vicious label, or even silence, or algorithmically delisted for coming forwards with your virtuous ideas and plans which may be in contradiction to emerging ideology. So doing the right thing requires real courage, yet many in a tech environment are too fearful and reliant on tech so are withdrawing.

These people who are withdrawing tragically are those with the integrity and insight to do the right thing, leaving those without the moral authority to thrive.

Face the challenge.

It is important to face what we don't want to face – for example, technology strategies may be fundamentally wrong and why that might be, so that we might do something about it.

We capture the issue and what good looks like in terms of:

• This book the psychology, corrupting technology language that is leading to the human obsession with and addiction to technology:

◦ to identify and open the debate around good technology and unhealthy technology

• *The Evil in Silence* - the pitfalls of how to prevent silence.

• *You Have Infinite Capability* - ISBN 978-1-9196337-2-5 and realising what good looks like

"The only people who make money out of IT are the IT companies".

I noticed that within executive roles in my 20s, IT implementations were screwing up a lot of businesses. One ERP implementation almost destroyed Lucas Wiring Systems. David, the CEO, went from 20st to being the skinniest person in the firm within a few months ?.

The additional admin burden of the ERP system shifted the focus away from established proven processes and created a data obsession, making data accuracy critical. Yet the simple legacy processes avoided most of these additional IT burdens even in very complex and responsive award-winning manufacturing environments.

When I was asked by *Management Today* in 2002 what best practice looks like, I said, "any systems should be centred on increasing the engagement of **people** in the process," i.e., IT / tech is very much secondary. Do not replace people or their value.

At this time, I noted a common mantra:

"The only people who make money out of IT are the IT companies."

Whilst mantras are flawed, that mantra has now materialised?

So why is tech so dominant? Is it really proven to be good for people or businesses?

What we now see has gone a lot further: many companies obsessed with technology are adopting tech strategies that are trashing their profits, differentiation, and employee well-being. So they are on a sure route and race to the bottom.

Tech - can you give it up, or are you addicted and duped?

We are heading for a monumental crisis. This is because our actions are fuelled by growth and domination so have become misaligned with the relationships/profit for the last 40 years.

This book explains how Tech/ social/digital businesses media manipulates, silences dissent or thinking, fuels misalignment, reduces profits, standardises products, and fuels more tech. We are trapped because of the well-publicised tech mantras such as these:

“Tech is the future” “AI you can trust” etc stops us thinking and closes the debate.

Tech is the antithesis of what good looks like for us as people, family members, and humans.

Since Socrates, our survival as human beings has depended on collaboration and maintaining relationships locally and globally. As a result, anything that interferes with relationships and profits needs to be heavily scrutinised.

The crisis/what crisis?

"The biggest crisis in the history of the world is coming. The biggest bubble in world history is getting bigger." said Robert Kiyosaki, bestselling author of Rich Dad Poor Dad *and famous financial market investor."*

And he still says more.

"The biggest crash in the history of the world is coming. The good news is that the best time to get rich is during a crash. The bad news is that the next crisis will be a long one. So, buy more gold, silver, and Bitcoin while you can. Take care."

Economic models are defunct.

The historical economic models are flawed. Tech is siphoning funds out of the productive economy with digital feudalism and economic rent.

So, let's look at technology

Because tech is limiting alternatives, tech companies are at risk of abusing a dominant market position.

So, what is the challenge? It is becoming aware of our own thinking to remove the stasis and open your options and start thinking:

The impact is being stuck in recruiting more of the same thinking focussed on growth and commonisation, when in fact we need to open the options and adopt a growth mindset to break the tech autocracy.

What really matters:

Our well-being really depends upon being acknowledged and valued for who we are by those around us and being allowed to be ourselves. The technology of technology is to assume that people are machines, yet our team, and wellbeing drivers are NOT data-driven they are very much driven by those around us, and we can make good decisions based upon minimal information. Notice how the excessive data required by machines dominate the debate.

Human wisdom and strategy very much depend upon our relationships with others and our ability to have a free and open discussion and being able to take responsibility for our work and act in our own way.

The technological environment is creating "dopes on a rope", with no time to think stuck in echo chambers and being fuelled by tech mantras. All the time during this process, people can't cope with the increasing speed, nor understand the technological outputs nor have the time to consider where this is going.

What is your core business?

Where is the clarity or critical thinking coming from?

Don't commoditise it, become aware of it and sell it, the price is what people are willing to pay, price is unrelated to convenience or cost.

How can you create healthy dialogue and new action?

There are 3 dimensions and skill requirements,

1. re-align to the challenge, remove obstacles,

2. Awareness and tools to transform capability on the job

3. Recruit for growth mindset and psychology to get the best from all.

Understand that you cannot measure performance in isolation and companies cause people to fail, not the other way around, people are hugely reliant on quality relationships and dialogue between themselves. This drives business success.

Get used to and value taking new action, what is the thinking that is causing blind spots and causing roadblocks?

The dialogue we have between ourselves in business is completely meaningless and valueless to ourselves, focus on relationships and profit, NOT growth and market domination.

It is tough to shift your perspective, take a new perspective and practice taking your own self-identified new actions and share the results with your peers.

You will not be able to compete with the FANGS,

Keep away from them, you have zero chance of competing. Can you remove yourself from technology?

Do you consent?

Rediscover what good looks like, value people for who they are, make time, listen be present.

Strong relationships enable us to do difficult things, or else when things get tough, relationships fall apart.

Empathy, getting to know each other, taking new action, supporting each other, creates grit and resilience.

Putting others first is a self-fulfilling scenario counterintuitively in dire circumstances it works best.

Beware of the negative generalisations “we can’t predict the future” and encourage critical thinking, if it quacks it’s a duck, if it seems wrong, it is wrong.

Yet do you notice with tech / social media the mantras are endless, so we carry on regardless for fear of missing out...!

A pre-pandemic view of what good looked like:

It’s possible to dupe teams by oversimplification too, such as rationalising all actions as to whether action fits with a customer focus. That’s not enough by far, we must take a broader view, which requires more sophisticated team working.

Post pandemic learnings.

Once the pandemic started, we started the "THE PUB". We used these 30 intense discussions to revise our perspective on the breadth of the challenge, which requires us to hit far more targets simultaneously. The impact of technology on well-being, loneliness, people losing themselves, fragility, siloed thinking, and suicide is blatantly obvious – so for example pt5. tech must STOP, counter to the panicky mantras of accelerating digital transformation which is squandering the priceless people – people touchpoints.

Overall, the concern is there is a lot of driving in the rear-view mirror, a lot of "doing" and very little "thinking".

Tech Companies are losing the skills to sell, leaving tech businesses unproven!

In the tech malaise, tech companies are destroying value and losing the ability to sell. With reduced transaction costs, they give away value or decouple their revenue model from their core business.

So, the growth of the business out of line with the core business does NOT equate to sustainable profits. So many tech companies when they grow without endless funding of losses will simply collapse.

For example, Starling Bank relied on interest on customer deposits and Revolut new card signup fees for profits, both sources of revenue are misaligned with their core businesses.

So, the precedent set having given away their core business means that when they start charging for services customers simply switch away from them, so their business models are unproven. In so doing they are trashing their own marketplace and sources of revenue. So a better mantra re "digital disruption" might be just "market destruction?"

To break out of this it's important to understand what type of psychology is driving you and your company strategy. Only then can you shift your perspective from which to take new action, or else you are stuck?

But do you have the skills to become aware of what is driving you and your actions!?

Shifting perspective without taking new action is worse than useless because it cements the status quo and stasis. It's vital to take the opportunity to take new action NOW. To understand your block, listen to your own internal dialogue in your head.

Tech is parasitic!

Watch out for sources of conflict, look for flawed assumptions, or those jumping or fuelling gravy trains. The PWC diagram below is based on old economic assumptions. The new technology might be driving pricing down and increasing consumption, but contrary to the past and tragically unlike previous industry breakthroughs it is NOT creating new jobs.

So, tech is reducing the size of markets and reducing the demand for labour, and quality of jobs and therefore reducing the purchasing power of consumers.

The nature of Tech sales intermediator businesses

Tech companies often intermediate the sale, so control the market, leaving the vendors with very little control. Interestingly, this effect caused Nike to start supplying direct. The collapse of Toys R Us whose digital sales became too reliant on outsourcing their digital route to market to Amazon was for sure related to their failure.

Sustainability of Tech Businesses – they are easy to replicate.

Tech is NOT difficult to replicate, so tech is not a source of top-line value or profit. Tech is undermining the source of profit, which is relationships, so tech is not even reducing profit. Tech is trashing profit!

Network effects make tech companies impossible to compete with too.

The overemphasis on scaling is a misnomer because many of the business models are broken or generate a negative margin, so, more sales sharply increase losses.

The cost of new customer recruitment without a sound relationship is also high and the cost of new customer recruitment increases too as these tech companies grow.

EQ vs IQ

The solution is simple but hard to identify, we are not looking in the right place. It's an EQ issue NOT an IQ issue,

Or…

Do you see how both statements create stasis and blind spots and are untrue?

What marketing activity drives your profits?

Here is a retailer we turned around…

When you know what drives your profits, and what is hard to replicate, align your message to what matters to your customers and SELL the value of your **core** business...

Retail sales has a journey, and it starts with positive association and relationships:

Contrary to the tech “give it away methodology” which undermines client relationships let alone profits, maintaining pricing matters - discounting your core business REDUCES total and UNIT sales too…

What are the mantras driving your pricing strategy?

A retail focus on discounts of core products **reduces** sales. You are just trashing your core business…

So, what is the issue to fix? It’s the psychology that is skewing your strategy.

Tech leads to commoditised products, take this example from a USA restaurant chain. A customer asked if they had blueberry muffins, the manager said no, but the customer could order the breakfast and just eat the blueberry muffins and chuck the rest away...

So, are you aware of the underlying thinking that is driving your strategy?

Commoditisation case study - selling online. Do you see how these mantras / generalisations trash profits?

And more growth does not imply more profits, this business had a negative gross margin so more growth = more losses. Amazon started making a profit when they shifted from vendor to become THE intermediator of the sale. In so doing, reducing their overheads, and increasing their control.

They are currently looking to control the logistics infrastructure regardless of where the sale originates.

The tech leadership Challenge

Where are the conflicts with the top 7, and why are tech strategies bad for our economies and social structures?

The simplification of tech products, such as the shift from inter to electric cars puts jobs at risk:

What are the mantras that blow up businesses?

Ease of doing business – the value paradox

The ease of doing business with Techcos also makes switching away easy!

So, what creates business loyalty is relationships and a positive association, NOT ADDICTION: yet addiction is probably the dominant factor driving tech sales?

Internally empowered staff create an engaging customer experience and profits. NOT the tech-driven notion of efficiency. Customer experience improves with delays in people's businesses processes because at these times customers get pampered by people. Empowerment creates grit in staff and motivates, internal staff motivation drives customer satisfaction.

Customer friction

The notion of customer friction (or delays) during a human-human contact-orientated sales process is value-adding. Friction is an opportunity to create profit and a positive experience, yet bots are used upfront, so drive customers mad.

It is baffling that women are not more engaged in AI/tech strategy, because they care more about people than men. Men care far more about things, not people...

Maintaining relationships

Maintaining relationships has nothing to do with technology. Text-based communication is a monologue, so there is no engagement, rapport, or conversation.

People should be at the fore of managing unexpected situations and decision-making, although not exclusively because we need easy stuff to remain healthy too. This broad view enables staff to maintain a personal and broad view of the process and customer experience.

Creating a place for everything reinforces accountability, process expertise, and learning. This is a people structure we put into an aircraft manufacturer.

The objectives below are conflicted which suggests the automation strategy is flawed: strong ethical values and AI/tech, alongside more emotional intelligence, might suggest technology is the wrong application?

People intuitively know if something is wrong…

Why will leaders fail and be duped?

There is no positive message in many tech businesses: their strategies create desperation, offer ever reducing, less engaging opportunities with no positive outcome, alongside an obvious negative impact on mental health and increased isolation.

Cyber risks

The cyber vulnerability risks are creating an escalating and impossible problem to fix, and more tech is NOT an answer. Yet reducing effective communication creates such massive blind spots and driving silence. So, the chance of critically thinking through the risks to get to the truth is negligible.

China will and are sitting and waiting to attack with their AI and quantum computing, making it increasingly easy to break encryption.

Tech and China

Chinese companies are controlled by the CCP and their bid to lead AI and quantum computing open the west to the unmanageable and escalating risk of a devastating cyber-attack.

China and Google really don't seem to care about employing people, the CEO of the Chinese giant, JD.com dreams of automating people out of everything and is partnering with Google.

Creating profits in the real world

The source of retail profits comes not from process and efficiency but from local relationships and knowledgeable STAFF: Take Ocado, the misnomers continue. “Efficiency drives customer experience”, nope, but it sure drives a dopamine addiction?

The tech financials don’t stack up either. Take Ocado: their automated tech operating model is more expensive than mixed bricks and mortar / online offer driving losses.

Yet what’s the advice from Accenture; the reason for the losses is because retailers have not embraced digital transformation!

Growth and domination case studies

ASOS the private equity fast-growth retail darling... the financial model is more expensive than a mixed brick and mortar model although the gross margin on apparel is higher than food.

So, growth again is driving this? Look at psychology, look at the impact on the environment?

Tech / AI simply drives sales and addiction, buying clothes of which 60% go straight to landfill.

This reduces market viability for players that are trying to do the right thing! Customers using online shopping need to vote with their feet and use the high street to maintain work and protect well-being and the environment.

An obsessive growth and domination focus is typical of a tech business and encouraged and valued by investors, over profits.

So, what is tech’s answer to the impossible problems tech is creating? Yes, “more tech”.More of the same and expecting a different result is Einstein’s definition of madness?

Why is this not getting raised and resolved? People who become atomised lose their sense of perspective, checks, and balances and echo chambers create a lot of doing and NO thinking.

China is automating to reduce reliance on people to mitigate the downside of those disenchanted with the scenario and the way they are treated. TheWest follows suit because they are being duped by the flawed growth and domination tech mantras. Yet the tech reduces retail profit, and the dominant Chinese tech businesses intentionally don't make a profit so, coupled with their weak social responsibility, are impossible to compete with!

Tech implications on the supply base and brands

So, what is the impact on the supply chain strategies? OEMs are taking back control of their sales pipeline from these aggressive online players that are destroying their margins and losing control.

Although musicians continue to suffer as their routes to market are increasingly controlled and businesses bullied by Big Tech.

So Big Tech is cauterizing innovation and dominating sales by controlling these markets. The FANGS are impossible to compete with and have a disregard for the commonwealth and privacy, so have become the "digital emissions" of the digital age.

The corporate lifespan is shortening due to this race to the bottom – falling profits, relentless focus on growth, although Amazon is far from safe, I expect Amazon could be trashed by the Chinese profitless growth digital giants.

Creating impossible problems

Misaligned actions create impossible problems. For example, people don't need much data to make good decisions, but tech does need a lot of data to operate at all, and our own data.

The techies have no idea how to tame the monsters they have created. It is nonsense to fix tech issues with more tech, do you see the stasis and blind spot created by the tech mantras?

The IQ trap – the ultimate unfair discriminatory tech outcome

The greatest unfairness is that you cannot train people to contribute to a tech economy below a certain IQ. So, AI will increasingly reduce work opportunities the more sophisticated it gets. Marginalising a vast swathe of the population from work will lead to violence. Families fall apart without work.

The role of governments

Governments are probably the only place where these issues converge nationally, so can be resolved.

Yet cryptocurrency/blockchain looks likely to shift fiscal control and voting out of their control in the new "atomised decentralised utopia."

Crypto has some interesting mantras driving it:

• Anti-establishment: claiming decentralised is more ethical than FIAT, bit coiners assume that through broader ownership of cryptocurrency good will prevail

• FIAT inflation and government control of money supply are bad

• You own crypto without central control, so have control

• Sustainability, usability, and security gaps will be resolved

• Visibility of the ledger with secure control of transactions

The issues are:

• People – people contact is critical to good decision-making, yet crypto is atomising people

• Individuals alone have no power so are vulnerable to tyranny

• National boundaries and national identities matter

• Someone needs to pick up the tab for supporting the poor

• Valuations are based upon flaws in our psychology, FOMO, and or investor hyped up pricing

• If crypto crashes, there is no fallback, increasingly digital means increasingly vulnerable to mass cyber-attack

• It's also the means for hyper surveillance and tax submission based on payments made at the microlevel

• The amount of data stored is ridiculous and data integrity issues overwhelming

• Power consumption driven by the ever-increasing security issues is unsustainable, bitcoin mining uses 50% of the UK's power consumption.

The Big Tech firms might sponsor and advocate destroying the current fabric of society and replacing it with what exactly? What do they know or care about society and people? Who will pick up the tab and support the disadvantaged or casualties of their systems?

The tech language is literally dangerously misleading and ultra-elitist or aspiring elitist. Here are some of the words from the IBM conference, describing their view, people doing more complex tasks creates some sort of hell, people need easy as well as challenging tasks to stay healthy.

Control of attention – the addiction trap

The tech systems are fuelled by controlling attention through addiction and triggering dopamine through likes, the buzz from purchases, and or causing distress in those being left out and forcing the stragglers through increasing traffic to need to follow suit.

The psychologist Brown said we should not trigger dopamine even if we can. Yet this is the tech world/malaise we increasingly operate in and are becoming:

- addicted

- the cause of people losing themselves

- like any addiction, psychologically fragile

- unable to cope with a significant challenge

...so more vulnerable to suicide.

The misnomer that people and machines can work in harmony.

The assumption that technology and people can harmoniously co-work is nonsense as the people and tech processes are like oil and water. The focus is on technology efficacy not our wellbeing.

How will we know what is going on? We won't! The notion that machines can define the truth or overcome bias too is nonsense. AI systems operate upon bias, so we are creating impossible problems.

Dated economic models.

The dated economic model above assumes that the old economic supply and demand models apply, but for the first time they don't. Technology is substantially simplifying products and reducing the labour required to produce them, as well as driving consumption!

This is the first time that innovation is reducing the net demand for new workers and skills.

It is important to understand that the technology is parasitic, and products are designed for simplicity to fit with their dominant models, such as Waymo self-driving cars.

Social media ad targeting

How effective is social media ad targeting, what is the ROI?

What is the personal cost to our sovereignty of shopping in a smart store, when you are necessarily under heavy surveillance?

Artificial intelligence

AI fuels the race to the bottom because products are increasingly commoditised, so price becomes the only differentiator, ultimately leading to trashing gross margins, products made and with zero people–people interaction.

Investor tech returns – until the pump and dump Ponzi schemes have taken hold VCs have not been making money either, so with so little else to invest in, have pumped up the valuations of growth and dominance consumer-focussed businesses, crypto, and fintech independent of profits. Funding rounds are at real risk of exposing investors to accusations of creating Ponzi schemes.

VC partners are not even looking at gross margin!

These tech business models are doing nothing to fuel productivity growth, they are feudal models that add no value, so make no difference to productivity. They do fuel top-end wage growth leaving those on low wages falling behind, and with less control and or opportunity to progress. Only 15% of people now get stuck in their first role nor manage to progress in their careers.

Are tech propositions compelling or just addictive?

66% of Fintech companies are simply not compelling. They are payment companies and like Amazon, Google, Apple are parasitic.

Digital is in fact far from capitalism, it is far more akin to Feudalism, gaining control over the flows of resources, and control of our attention.

Digital businesses are rent and power-seeking. Social media is integral but avoidable. However, instant gratification makes it addictive. So have we lost control of our actions and our mood?

Big Tech buy and build

The Big Techcos control innovation, because they are unassailable and iteratively gobble up potentially competitive players that might keep the markets healthy.

The digital models create intermediator-enabled products. These can drive multi-asset usages such as ridesharing, and bike hire.

Unlike the introduction of tractors for farmers, rather than creating new industries, such as making tractors, for the first time ever this wave of Technology “perceived as progress” is in fact disastrous because these tech industries are fundamentally reducing work, the meaning of work, and reducing profits for the productive parts of the economy, so thereby shrinking markets.

This enables higher utilisation of expensive assets and leveraging value by moving from ownership to rental. 30-40% of employees drive for a living, so autonomous cars are bad news for them!

Digital competitiveness.

What are the dimensions of digital competitiveness? Who do digital companies serve: the protection of themselves or what is best for people, broader business interests, and society?

Companies like ASOS are more vertically integrated vs. a sales intermediator like Amazon, so have far less chance of making profits. The market power Amazon exerts enables them to drop suppliers as they see fit without repercussion.

FANG companies focus on control of resources.

The FANG companies are far less about competition and far more about control of and controlling the flow of resources. The Apple service model increases their power over their user's media usage and competes on price, ultimately undermining their supply base.

The perverse situation is that this power affords the intermediator of the sale far more profit than the business providing the goods!

Big Tech entrepreneurs could be described as unproductive, who buy up rivals to stifle competition, thriving like parasites on brands and the productive parts of the economy.

Economic rent

The work that is created by these tech intermediators is unengaging, lacks a compelling purpose, and is diminishing, so what opportunity does this provide for the atomised next generation and or creative types?

Work shifts are a misnomer.

Only the hardest tasks will be left to do for people. This will marginalise all except the very brightest. Purely hard tasks create stressful and unpleasant work. People need simpler tasks too, to create a sense of accomplishment. A people–people interaction experience matters to wellbeing.

Yet the flawed mantras at IBM continue. Dumping the bottom quartiles of the workforce is hardly a venerable aspiration, and “mundane” work, might be routine but does not have to be mundane and is a useful data gathering and learning opportunity for people?

The big 4 tend to jump on gravy trains and structure their services around extracting funds between the sources and intended recipients, in government and tech. This is hardly impartial advice.

Who will be left behind – YOU!

The bottom 40% of the population will be left behind, with less access to knowledge and skills that would enable them to participate in work. All work is at risk, we simply won’t be able to train fast enough. These skills gaps are currently driving the focus on faster automation!

Social media fuels division enables echo chambers, depression, anxiety, and polarised views. It is important to understand you are addicted to these platforms, and like any drug, they probably have negligible value to YOU. They waste a lot of time and create a lot of harm.

The impact of social media

The importance of work to our wellbeing

The impact of not having work leaves many feeling worthless and undermines the traditional male role in the family. Work is the backbone of families, and families the backbone of society, so ANY work should be preserved and created at all costs.

As work is critical to society, high value or low value, the notion of workless tech utopia is an absolute fallacy. The reality of worklessness is despair, suicide, drug addiction. Creating work is a critical role a business contributes to society. No one will pay you to do nothing either, so you will probably become tech slaves of some type.

High-margin SMEs are the backbone of economies, not these Techcos, yet the Chinese intermediator Alibaba is listed in the USA and moving into b2b sales in the USA, potentially capturing control of the US economy!

“The next Amazon” – the lie

The aspiration of a start-up becoming a FANG company, or the mantra of “becoming the next Amazon “is completely unrealistic, Price's Law suggests the probability is 1x10-24 roughly zero!

The notion that tech will follow the logic of the past where farmers buy tractors and a whole industry opens around tractors, is flawed. Tech is NOT creating a follow-on industry in this way; it is destroying the productive parts of the economy.

The tech reduction in GDP

People's unique capabilities are being squandered

People have a unique ability over machines to think and envision the trends and anticipate the future. This does not fit with Big Tech or China's interests because the tech actions converge on dystopia and are NOT inevitable, so it seems attempts are being made to control and keep people's attention in the present, without the bandwidth to consider the future or learn from the tech or other failures of the past.

People are machines is a common, convenient and false assumption. Tech is unable to allow for the uniqueness of human behaviour, so the assumption that we are homogenous, or machines, is convenient to try and model. Yet people are far from being predictable, so the assumption that people's behaviour can be modelled on a few factors is nonsense. Medical diagnoses using AI requires knowing the full picture, so is hugely complex too, and to what end anyway?

The techies have so little knowledge of people too, and in fact, the best psychologists remain baffled by many of the emerging people-related factors.

Implementing technology as part of education is tragic because it can change the form of a young brain and cruelly reduces people-people interaction when they need it. Learning is highly dependent upon relationships and environment, children's learning very much relies on the environmental aspects and close and safe p2p relationships with their peers, which as we know tech destroys and atomises individuals from others by plugging them into the tech network.

Their emerging wisdom is that our minds are being rewired to think less, look forwards less and plan forwards less, dope on a rope.

Where are the value and learning if we are not thinking nor allowed to take and learn from the actions we take in our own unique way?

The tech mantras

Generalisations or mantras such as "people are more adaptable than machines" drive strategies and create blind spots. Government leaders should be more careful and aware of the implications of their words.

Tech is creating very boring work.

Healthy dialogue is critical to reach the truth and in so doing encourage critical thinking. This is the means to evaluate risk and anticipate possible outcomes.

Yet the tech world assumes that “tech is inevitable”, “AI for good” - these mantras close the debate, dumbing employees' capabilities, and corrupting them, by trapping them in a “tyranny of tech tolerance.”

Who can we trust, or should we trust, to develop AI? Whose interests do the providers have in mind?

What happens when we trust Techcos more than our governments? Who will pick up the pieces or look after our national or local interests? Or will the dopes on the rope get blown around in their atomised states of isolation from others and integration into the tech malaise?

These mantras such as tech is inevitable, drive strategy. This happens without thinking or planning and 42% are just carrying on without a business case.

In retail, there is a lot of “nonsensical doing and little thinking.” Yet retail profits are driven by people – people relationships and positive brand association, not convenience and fast impersonal service.

The tech obsession with data is not the fuel for the human machine. Data is the fuel for the tech machine. It also drives the network effect to increase monopoly power.

So, notice the obsessive discussion about data in tech and broader business forums!

Anti-competitiveness is now more about low-priced dominant players not dominant players overcharging.

Giving away free products is not clever and is anticompetitive.

Zombie techs are trashing (their own) markets!

Companies like Revolut have uncompelling products, give away services, and so trash their own markets. This undermines the productive parts of the financial infrastructure, fuelling losses.

Starling Bank, unlike Revolut, is an interesting business with quite a good proposition. It is compelling although its revenue is not linked to its core business, so when the business scales the core value proposition is undermined, not proven nor robust, and so vulnerable to huge losses and collapse.

The creation of weird jobs and outcomes

Because it is doing a lot of questionable things it's not surprising that the tech mantras and language fuel some weird job titles and outcomes, and leave some blatant gaps unchecked… creating blind spots! The three manager roles on the bottom LHS are annotated to the original cartoon!

The final blow, exponential speed – people simply can't cope with it A leader's fear of being left behind overrides their rational decision-making and doing the right thing. This is driven by exponential increases in speed as the chart shows since the late 2000s.

If social media platforms are designed to control your attention, do the social media platforms have your best interests at heart? Or are they just fuelling addictive disorders? There is little willingness to provide visibility on how these social media platforms operate.

The role of healthy debate

The tech mantras are difficult for employees to challenge or disagree with, so create blind spots with areas of risk remaining unchallenged.

When these blind spots are challenged externally, every effort will be made to silence critics or label them as Luddites or having grey hair, see *The Evil of Silence,* ISBN 978-1-9196337-0-1. These labels create polarisation in views, echo chambers and stifle debate. Because there is a cover-up anger ensues because the tech arguments are often insubstantial, indefensible, or based upon flawed assumptions or motivations.

How do employees feel?

37% feel that their work is NOT meaningful? Where is the meaning or engaging purpose in a tech business, save maybe a rebellious mission and ill-thought-through narrow solution? Increasingly workers at the bottom end who dominate such as Amazon employees have fewer and fewer rights or representation.

Leadership skills are defunct – but this is unseen.

Leadership dialogue is weak, here is the analysis of a successful leader's communication, most is vague and general. This is the type of dialogue that creates stasis and silence.

Jordan Peterson finds that most of our thoughts are wrong so our thoughts must be continuously challenged by those around us to stay healthy. So it's so important to become aware of weak dialogue. A specific example of the underlying psychology/mantra that drives tech businesses is "people are machines", so, what if people are not machines?

Every Utopia reveals the personality of its author.

Jeff Bezos, since childhood, has been obsessed with space and science fiction and fears stasis.

Daniel, CEO of Satalia, says “people are machines”, so explores how can we get a utopian world where we can operate as a swarm, decide each other’s earnings as well as meet the needs of those that inevitably find themselves without work.

Mega Trends

Ridesharing is a megatrend, and fleets will be autonomous. Whose interests are smart cities in?

So, in conclusion:

You are just doing and looking for things that fit with your strategies or driving in your rear-view mirror unchallenged.

What are the common mantras in your business, are you aware of them or too familiar to see them, do you have the skills to identify them?

What drives you and how self-aware are you?

The paradox is “growth and domination” obsession is a relic from the recent past, not the NEW NOW. So must be dropped. Disagree, how does growth fit in with the obvious casualty of growth sustainability?!

What matters in the new now has been highlighted from the pandemic, profit, and relationships.

Profit gives us freedom and relationships are essential for wellbeing and profit – it is circular.

The challenge that remains is despite charging away with technology that we are hugely unaware of what makes us tick and or know much about people at all.

Yet we are surprised when there has been a substantial drop in wellbeing when mental illness is clearly caused by technology.

It's important to understand what AI/tech is: it's a parasitic service that undermines/consumes its host or the productive parts of the economy.

Tech is not a means to an end, apart from a weirdly naive outcome for the Techs. Tech for sure is not good for people planet profits and undermines our ability to get back into control!

We must encourage free speech, open the options, and encourage critical thinking to regain balance, wellbeing, and sanity.

Why must we act to reverse these trends? Look at the madness that is happening all around us and the increasingly obvious havoc we are wreaking on our planet due to our obsession with growth and domination fuelled by technology.

Why will leaders fail?

You will fail because you have inadvertently silenced your employees. You are trapped in your tech-driven thinking and are unaware of the warning signs, what drives you, so your self-awareness will be low. Therefore, you have little idea of what specifically to fix to regain control.

Wars start when the talking stops.

Tech mantras hide the truth, but when you bravely open the dialogue and focus on the top 7 you will start winning, undo the tech cancer, and begin to realise and regain what good looks like.

Relationships are the lifeblood of businesses. When Provident Financial took out their sales agents, the business crashed.

As things will get much tougher, the importance of robust relationships, open dialogue and immediate action, and empathy increase exponentially. Right now, what we have is fear, major blind spots, and the wrong people silenced.

Recognise too where you are losing control to Big Tech, then focus on doing what is best for your business.

Winningthinking® perspectives are required to shift thinking and take new actions. There are 3 main aspects, tools to realign to the challenge, tools, and psychology to transform capability, and recruit for growth and the right psychology to get the best from all. The number of immediately self-realised actions taken is the best indicator of the success of this structure.

There are no skills shortages, people can learn vocationally, and you have plenty of capable people.

What to do?

The actions to respond to the challenge then are threefold:

1. create the tools to face the challenge - open the dialogue,

2. the way of working to transform capability and

3. recruit for growth and psychology to get the best from all of your people.

Rationalise every tech and business proposal vs does this improve people–people relationships?

Or am I still obsessed with growth and domination?

In the meantime, can you and your team have difficult discussions or are you avoiding them?

Who has the courage to lead these changes?

Develop robust people–people relationships so you can manage and keep the right people on board to face and see your business through the hard times.

You have been wrong for 40 years - be humble!

Face the reality of tech, you are losing yourself, you can't cope with the speed, people don't need the data, and face the devastating impact on your people's meaning and work and chaos all around

– maybe your tech plan/strategy is wrong!

Have you lost sight of meaning in your business, does it mean anything meaningful to anyone?

Find the humility to face the fact that you have been wrong for a long time, taking the wrong actions and your skills to navigate and cope are largely defunct.

Can you see beyond the short-term financial issues?

and do you have a team that can move on all fronts to hit the right target?

Encourage the people who care – you've trashed them.

Do we live in a (tech-fuelled) machine that is designed to neglect what is important about life?

What is important about life?

A Jazz band is what good looks like?

A Jazz band (The Ukulele Orchestra of Great Britain)

What are the components of a Jazz Band?

- Intimate relationships
- Oneness
- Acknowledged for who you are
- Valued for your unique value and view
- A job – work
- Profit and relationship led profits

Don't underestimate the challenge, you and your team are probably addicted to technology – you don't have control so tech should be legislated against?

"**Addiction** is an inability to stop using a substance or engaging in a behaviour even though it is causing psychological and physical harm." (Medical News)

Start thinking forwards and asking the tough questions.

What happens when the Megatech companies, that employ no one, start fighting/competing using AI?

The stakes are high.

The dangers of tech are increasingly impossible to defend against. The more you digitise, the only way to reduce risk is to de-digitise. This battle hasn't even started. China are sitting and waiting

- will dominate and present a very real threat with their leadership of AI and quantum computing.

The UK Government proposes a response to a significant cyber-attack is nuclear weapons.

What can you do? Check your actions vs impact on relationships and profit.

Finally, supporting local businesses – local social media Facebook and similar networks are run by completely untrained unaccountable people responsible for vetting content, yet people's local businesses depend upon these platforms. It's time to upskill to navigate this scenario.

Action: *You Have Infinite Capability,* ISBN 978-1-9196337-2-5, explains the challenge and solution to prompt self-realised action to make these shifts.

No new paradigm, no new action – no new action – stasis.

Your call.

Tech triggers dopamine like alcohol or any other drug addiction, so it's an addiction, which means you have no control over your actions. The incentive for these competing platforms is to increase "watch time," which of course is control of your attention and facilitated by the most addictive platform.

We need to bring the algorithms and tech strategies out into the light to raise the awareness and visibility of the issue to resolve it.

Equality, Diversity, and Inclusion the: pathology, the destruction and lies. Achieving business restoration, and THE cure

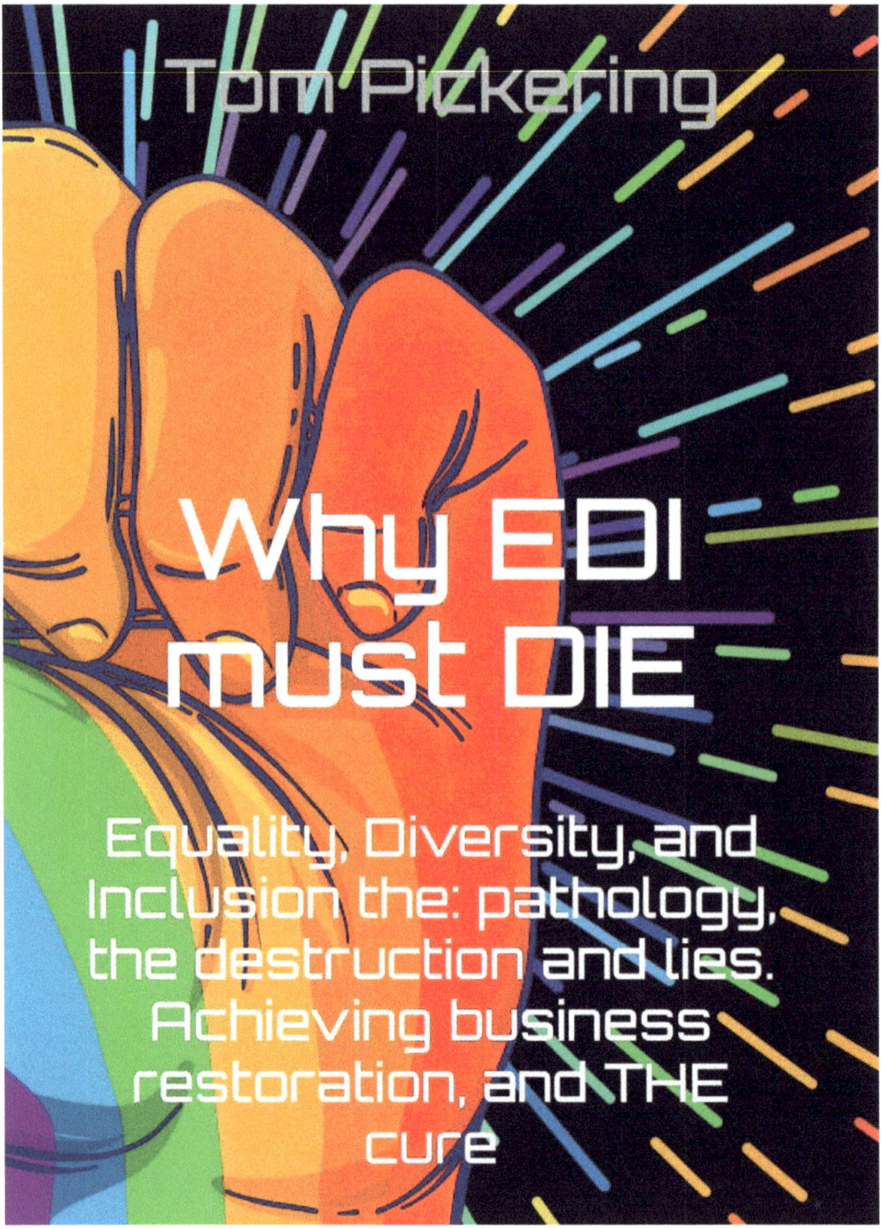

To save your business I need you to understand why EDI (Equity, Diversity and Inclusion) must DIE.

"Egalitarianism is their excuse for failed elitism." Peter Thiel

Equality, Diversity, and Inclusion the: pathology, destruction, restoration, and cure This book sits alongside these books because alongside the pathology of silencing and tech, and not being able to stop, I realised that we are unknowingly trapped in an EDI nightmare and unable to get out:

How the "Woke" is destroying our world and

your business (and what to do about it)

1 Thessalonians 5:3, NIV: While people are saying, 'Peace and safety,' destruction will come on them suddenly, as labour pains on a pregnant woman, and they will not escape.

"Peace and safety" are the divisive guises that underpin EDI. Even The Bible suggests destruction is coming on them deploying these strategies.

"Implicit in digitalisation is standardisation and universality and becoming like our enemies to mitigate the downside."

Peter Thiel adds "Egalitarianism's excuse for failed elitism."

In 1994 I was a director of an auto manufacturing company and I noticed that it required real skill to be honest. The CEO was great at drawing generalisations from isolated incidents to beat up his executives and create a smoke screen for lack of process.

Wokeness, according to Eric Kaufman, is the sacralisation of race, gender and sexual identities, so if any way you speak out against these groups of the ideology, you have committed blasphemy centred around social cultural socialism: equal outcomes and psychological harm protection.

People are very open to having their attitudes shifted in their late teens and early twenties, so these attitudes will likely solidify, and so social culturalism is only getting started. So pay attention.

The truth is, Wokies are destroying everything. Your business. Our world. The economy. Our children's futures.

So if you're ready to get a grip and ditch the slave-minded shackles that have silenced too many CEOs (and their employees) for too long, then keep reading.

It's time to:

• question reality and reject corruption

• stop emotionally reacting to something you don't understand for fear of being outcast

or labelled

• learn how to read situations and identify blind spots and manipulation

• challenge the weird dialogue

• ask the questions no one else has the courage to

• and discover the real solution to fixing the woke fallout.

In short, it's time to seek the truth. Woke is the parasite, not the solution.

And the enemy has a manipulative plan to silence us into inaction. But not everyone wants to be silent. We need you.

If you're ready to start a much-needed dialogue into woke and stop its crushing ripple effect on your life, buy this book today.

Divide, destroy, and entrap without firing a shot.

I'm witnessing enough weird conversations and crashing outcomes to indicate CEOs realise that they might be out of their depth.

In part, it's due to “slave-mindedness”, which is rather like getting drunk and falling asleep on a train, rendering you unable to get off until the guard wakes you up at the final destination.

The challenge for many leaders now is how to regain the ability to get a grip and arrive at a destination of your choice.

Your perspective matters too: “Set your affection on things above, not on things on the earth.” Colossians 3:2 KJV

Christian principles produced a successful civilisation. So what do you now believe and what does good look like?

Take men: we are the gatekeepers of morality; we are the warriors who stand up against corruption and protect women and children. So, it’s convenient for those who wish to create problems to label us as “toxic” to undermine the male role in averting corruption to keep society safe. This also undermines the male/female relationship that’s the crux of society - the nuclear family. The ultimate attack.

So what is the purpose we are supposed to be upholding?

The Economist in Nov 2022 reported ESG/diversity progressive stocks have not performed, and in 2022 reports Woke Capitalism is creating a tyranny

So - What is the purpose you decide on, is it robust and does it hold water, are you acting or just virtue signalling?

Economist Nov 2022 showed Human connection, intimacy and empathy now really matters.

And what is being attacked? Take our identity and history: Winston Churchill is a symbol of defiance in the face of tyranny so an obvious target for those who advocate tyranny. As Kennan said, “We must have courage and self-confidence to cling to our own methods and conceptions of society. We don’t have to fight. All we need to do is remember who we are.”

And people of faith practice gratitude. We are the luckiest people in the history of the world. Yet the health and safety focus suggest that we are at risk of being "unsafe," the perquisites come with more tension and threats of cancelling, compelled speech: pronouns than they solve. Targets deemed to have unsafe trigger warnings, ranging from William Shakespeare to the terrifying *Little House on the Prairie*. In creating such havens from harm, universities, and organisations pander to the most debilitating levels of emotional instability. John Cleese suggests that we should not organise society around the most upset people, else we will create the most neurotic type of society, isolating people from learning, new experience and new action in order to build the grit to cope with normal life.

We should not trust our senses.

Class used to be the main issue, but now the working-class are the most marginalised because other groups are more emotive, dynamic, alternative, identity-centric targets: women's studies, black studies, queer studies, whiteness studies from which the new radicals sprang up.

The death of George Floyd provided an excuse for the radicals to signal panic, racism is systemic, unlearning racism leads to calls to abolish whiteness.

We are now told who we are, and we must accept it, working to abolish men, leaving the working-class contingent most likely to be desperate, isolated and commit suicide.

John O'Sullivan states communism isn't oppression; it's oppression, plus preaching a doctrine that you know to be false.

Yet the solution is to search out the truth and think for us and replace the mantra of compassion with excellence.

Yet our culture of liberty and national tradition is being substantially deformed out of recognition by critical race theory and its attacks on our national heritage.

Conservatives are uncharacteristically silent, running scared of the media, yet this culture and protecting what is good are the main reason for being a conservative. Ofsted doesn't enforce what is in law already in that you are NOT allowed to politically indoctrinate, the same requirement holds equally for big tech algorithms.

Information is biased, 7/10 students had not heard of Mao, and the anti-work conservative side is dramatically underpowered, so it is time to upskill. Policies won't work because the ideologues don't see the issue, they have lost sight of what God looks like in their morass of inhuman activism.

Solzhenitsyn says that the most accessible key to our liberation is simple "a non-participation in lies".

As I have outlined in my previous books, combined tech and work factors have created a mass formation psychosis or a collective hysteria. Hysteria takes root in tough times because people are looking for a narrative to latch on to to normalise their status quo, give them something to fight for and give them purpose.

The Archbishop of Canterbury said, "Why is it such an agony to remove a memorial to slavery?"

The dialogue is misleading because this wasn't true. The Tobias Rustat statue wasn't a memorial to slavery. It was a memorial to a benefactor, so all our history is not about race and slavery, yet no matter the colour of your skin, you must go against your experience.

Compelled speech such as pronouns, are a dangerous thing because it misaligns your actions and beliefs, compelling you to say something that you know isn't true or speak out against your conscience. It's an assault on your integrity.

Yet leaders are hounded out for refusing to call a transgender woman a "she", which is scientifically inaccurate. The message this sends is that we should not trust our senses or our own reason.

This is gaslighting and narcissistic. A society of people who don't trust themselves must therefore be infantile and servile to an ideology.

The march or mass psychosis in 1930's Nazi Germany got to the stage where people said, "this has gone too far, it's helpless".

We must recognise and openly discuss to check and search out what is true, beautiful, and good.

This is a choice between spiritual independence and servility on all fronts.

*(*1 in part this chapter is a precis and commentary of the essays in Fighting Back published by the*

New Culture Forum)

The importance of national identity

As we can see in Ukraine, we are witnessing the birth of a nation when there wasn't a nation there before.

We should see the rebirth of patriotism because that rightfully recognises the achievements of the pat because the opposite is some form of universalism.

A sense of nationhood remains central, and a state and its people need to be grounded in national identity. Our identity is rooted in a sense of place and circumstance, not a set of universal values without any roots. So, the disaster of liberalism is its claim to be founded on universal values when there are no such thing as universal values.

Values are narrowly, culturally, and nationally defined, rooted in a place and historical experience.

The French Revolution put everything in tens because that was rational, yet there are 365 days in the year!

So, society is not a contract for fragmented individuals, it is a contract that spans across generations, past, present and future, the living and with a responsibility to those yet to be born.

So, a universal set of values in a single system of international or even EU law is an absurdity.

Even though the EU is fundamentally made up of remarkably different nations, it is not culturally homogeneous, or else we are heading to the disaster that we face now, or the disastrous outcome of the French Revolution.

The current world looks horribly like George Orwell envisioned in *1984*.

The Germans have not come to terms with their history either, which may explain their closeness to China and Russia?

The Bible acknowledges the importance of nations and maintaining national boundaries too.

The American Declaration states, “We hold these truths that all men are created equal”, yet this statement is deranged, even when applied to the now superior sex of women.

We are born wildly different, utterly dependent upon each other, and not free. As St Augustine bluntly said born midst “piss and shit”.

The UK has only been a democracy for 100 years or so, too, so we are quite young on the journey.

The Roman empire was a single universal international empire that met its absolute collapse in 410.

The UK is not a product of enlightenment; its structure was propagated by worldwide expansionism, trans rights, the toxicity of racism and other slurs.

So liberty and freedom are not universal, and it doesn’t export very easily as neo-conservative efforts in Afghanistan have shown to impose woke sexual values on an Islamic tribal society, so even soldiers were seduced by this work ideology.

The family and role of men and women are the crux of society, undercut by identity politics.

Universal human rights were originally designed to protect the individual against the state, but when they are directed to diverse minority groups, they strengthen the state against the individual.

We need to consider West is best, based upon our successful history of opposites being reconciled and stop intersectional policies that identify oppressor and oppressed. Lauding the victim and despising the victors is a catastrophe for any sort of progression, sensible analysis or achievement.

So, our values and purpose are rooted in real history, not doctrines, so imposing One Love armbands upon the Qatar World Cup is absurd and arrogant.

Our history, freedom, and liberty are the basis of our identity and the state, as set out by our ancestors.

You will not find them in Qatar, China, or Russia but through a process of acculturation.

So as David Starkey says, we must regain a sense of patriotism and gratefulness for our past, alongside humility, because we did nothing to create it, we are merely the inheritors, but the blame will be ours if we let our inheritance slip through our fingers.

*(*1 in part this chapter is a precis and commentary of the essays in Fighting Back published by the New Culture Forum)*

A faith perspective – CRT is a complex web of lies and anti-Christian ideology.

In Acts we learn that we are One Nation united under Christ…

The Church of England declared itself institutionally racist. Its no longer good enough not to be racist; one must be anti-racist. The Church is infested with the critical race theory ideology. This is absurd because the notion that all white people are one homogeneous group, think alike, and white are privileged is a false racist idea in itself.

The ideological lie that underpins this is that colour is a personality trait, so if black people don't feel oppressed, they are accused of perpetuating whiteness.

Biblically, this undermines the basis of Christianity because the colour of one's skin is therefore a sin you cannot escape. There is no escaping the sin of whiteness, so there is no salvation, yet a person of colour cannot be racist in the eyes of CRT.

This is the mode of entrapment in groups dividing people based on immutable characteristics.

It would be more beneficial to look at ourselves as British, which unites us all, or even better, find our identity in Christ.

This ideology survives under the umbrella of "This is my truth", where anecdotal experience trumps truth and fact, hence the accusation of lived experience, as a white person, you would not understand etc.

For Christians, the truth is Jesus Christ, so in essence, woke ideology is fundamentally anti- Christ, yet the Archbishop of Canterbury declared at the Synod that the church is institutionally racist and must repent!

The outcome of following such ideology is getting corrupted by getting drawn deeper into this divisive language. Declarations of positive discrimination too are misleading. Discrimination is discrimination and co-opting people into governing bodies that haven't been democratically elected or motivated by this ideology will create distrust, not merit.

Shortlists are corrupted by unconscious bias training, ideas that do not work, and deepening division. This division can be leveraged, A pastor claimed that those following captain Tom supported a cult of white nationalism., divisively undermining the basis of well-being, nationality and identity in one go.

So, if everything is seen as racist, family values, marriage and Christian teaching are side lined.

Queer theory creates convoluted lies to cover the fundamental lies, so in the case of the Church putting the individual before God, or business interests as the case maybe, this thereby undermines the first commandment and or the profits or meritocratic basis of a business.

Symptoms of division run deep, re baptisms for people who change gender, and homosexual Church marriage, when the Biblical sacrament is that a man marries a woman and loves themselves.

as they are, not pushed into surgery to mutilate their bodies. This leads to Bishops openly having sexual relations outside marriage, creating an anti-Christian agenda that corrupts Biblical teaching and which is not rooted in Christ.

The Church Report contested heritage gives instructions on the best way to remove statues that have links to the British Empire, erasing history, furthering anti-whiteness. The idea that any monument should be of someone perfect is anti-Christ because it denies that we are ALL fallen.

So, we are in a spiritual war for the soul of the nation. Jesus did not come to bring peace. He came to bring a sword.

The notion of peace and safety, is demonic:

1 Thessalonians 5:3, NIV: While people are saying, 'Peace and safety,' destruction will come on them suddenly, as labour pains on a pregnant woman, and they will not escape.

The claims of ideology "undermining my safety", therefore, is another powerful falsehood and negative generalisation. Negative generalisations are always false.

*(*1 in part this chapter is a precis and commentary of the essays in Fighting Back published by the New Culture Forum)*

"When in Rome (do as the Romans do)" idiom SAYING when you are visiting another place, you should follow the customs of the people in that place (Source Cambridge Dictionary)

Mass migration would ensure Labour Party election victories for years to come. In 2000, the UK was opened to mass immigration because ethnic minorities voted overwhelmingly for Labour: 82% of black people in 1997, vs 12% voting conservative. Take London; most Londoners are multiethnic and non-white. 82% of Muslims think of themselves as Muslims first rather than British, 15% of Muslims empathise with extremists, 4% support terrorists and suicide bombers: 4% = 120,000 people in the UK.

Rod Liddle notes that when Muslims were complaining about the LGBT training of their children, the BBC did not know who to support!

Young British-born Muslims have far more moderate views and regard themselves as part of British society, yet a significant minority hold far more extreme views than their parents; 36% believe a Muslim should be put to death if they leave the Muslim Faith.

So, in communities such as Blackburn, mosques are run by Deobandis, the group that gave birth to the Taliban. Poverty, unemployment, and marginalisation make these places a seedbed for cultural practices, honour killings, grooming gangs.

Yet petrified by political correctness and the mantra of ant-racism and multiculturalism, councils have remained silent.

This leaves a rebellion against wokeness homophobia and misogyny on the increase.

Theresa May allowed the rise of 85 Sharia Courts in the UK too.

Failed by the government to define what it is to be British and uphold standards, what British model do the young have to aspire to or follow? The model is being denigrated legacy and achievements, and pupils leave school leave without any knowledge about our proud history or who Winston Churchill was!

So, without having a patriotic view to aspire to and encourage, integration is a nonsense. In fact, more likely to deepen sectarianism and unrest?

This requires statesmanship of herculean courage: courage requires new action, new action requires taking new perspectives, and only then will communities appreciate the value of the British way and outcome.

So fundamentally changing the ethnic makeup of the UK changes our culture. So wokeness must stop trashing our culture and heritage, or it's likely we will have rebellion and unwellness.

To protect their national identity, the Danish insist that not more than 30% of their community should be non-Western immigrants, and they promote the Danish culture, history, and patriotism.

The uncontrolled mass migration unleashed by Tony Blair was arguably the greatest act of national self-harm in the history of the UK, a reckless and dangerous decision. Conservatives have supported this policy too.

The media

The main objectors to the anti-Rwanda policy to deport illegal immigrants are open borders fanatics. Yet, if you cannot control who lives in a country, what does nationhood amount to?

The BBC/media must allow consensus to be challenged to avoid extremely costly policy decisions being made. It is the BBC's duty to challenge. The BBC was in favour of Remain until the British people said they wanted to put an end to the misleading EU project. John Bercow was supported by and applauded by the BBC.

Refuting bias is difficult in media faced with the mantra of "well, that's your view".

The *Today* programme made it clear why Brexit would not work and made little or no mention of the benefits.

The BBC is policed by Ofcom, yet Ofcom is a retirement home for former BBC employees such as the Chair of the BBC Michael Grade who became Chair of Ofcom.

50% of people no longer trust the BBC to be impartial.

*(*1 in part this chapter is a precis and commentary of the essays in Fighting Back published by the New Culture Forum)*

Putin believed the West to be decadent and weak, but his actions in Ukraine showed the power of nationhood and patriotism and identity in Ukraine and the corruption and weakness of his own regime.

The West tends not to have a long-term strategic plan like China but innovates creatively and is never happy with the status quo.

The crux of the West is not an Xi like God-king, it’s one law for all and liberty under it, releasing energy from below rather than a regime or strong man imposing ideology.

C20 shows the cost of standing up against the Nazi and Soviet tyranny. Yet, the anti-Western intellectual culture depends upon Western decline, some looking at communism inevitably winning over the West, as Xi perpetrates now under his resonating tyrannical mantra “Chaos vs Order.”

All democracies paid the price for letting in tyranny.

Regan, in 1982, said: “What type or people do we think we are? Free people worthy of freedom?”

The West is easy to underestimate due to its openness. But our freedom, history, system, and patriotism make us stronger. Yet our enemies believe that the freedom we cherish is weak and decadent.

The Poet Goethe said,

“That which thy fathers bequeathed thee.

Earn it anew if thou wouldst possess it.”

*(*1 in part this chapter is a precis and commentary of the essays in Fighting Back published by the New Culture Forum)*

Yet if men act as men now, calling out the falsehoods of EDI that are persecuting them and undermining their role, they are labelled as bad people.

We must get back to accepting people for who they are, valuing different views with a factual sense of identity, rooting, and rejecting corruption to allow us to re-align actions and make good decisions.

The Bible suggests there are three types of people: the wise, the foolish and the evil.

So, from where do we derive our wisdom?

Then there is the rise of Sam Bankman-Fried (SBF), the clever CEO without wisdom.

They get found out.

(Source: *The Economist* November 2022)

It's reassuring that good always prevails over evil because love is the only indestructible facet of life.

Right now, a whole raft of misaligned dialogue and ideology is intentionally creating misaligned actions with reality.

Yet there is a vast opportunity to bring businesses and people to life.

"Innovation has stood still for the last 10 years or so, so with our increased wealth, what have we done?" Peter Theil.

Good independent thinking and ideas now don't get traction. It's because people have become slave-minded, and most are trapped in "the matrix", so losing their ability to think for themselves.

But the truth always finds out corruption and deception:

Beveridge, in 1942, during WW2, wrote, “a revolutionary moment in the world's history is a time for revolutions not for patching”.

This paper is about reading situations and identifying blind spots by following the manipulation and understanding the underlying drivers. Knowing who the enemy is and where it is leading.

Listening to what you are expected to do and understanding the outcome and tools needed to get a grip on your situation.

The alternative is being slave-minded - emotionally reacting to something you don’t understand without understanding how you feel or why they feel passionate about what they feel.

Pay attention. You are attacking fellow man, not the cause’s elite. To say nothing without punishing yourself is godless and without shame, learning or realignment.

The protagonists need a virtuous idea to hide behind, from which to hate and attack intolerant ideas of their ideas. Take the popular EDI dogma again, “you should look different”, but think the same and follow the same insane dogma.

The bigger picture is keeping and protecting power, as you would if you were in the elite at all costs, by controlling currencies, spending, and social scoring.

The only issue holding this policing back is the capability of AI because the current policing systems are not scalable. So maybe you are feeding that aspect of the emerging tyranny too. Is AI now a source of wisdom?

Blueprints have changed too, for example, the model of owning property. If it can be taken off you, you do not own it, or in the case of China and Russia, if you cannot say no, you have lost control and need to exit/distance relations. Tough decisions, yet the German Government digs their businesses deeper with China.

If you delete one side of the argument, you never get the truth and create slave minds. Slave minds ignore truth, believe in nonsense, and even evil ideas, which hijack your competence, so you are on the train or in the matrix.

It is vital to have the skills to discuss and disagree – without individuals being singled out and censored.

"In every coalition, there is also some co-loathing," Yitzhak Rabin.

So, what do you believe in?

I did a turnaround talk two years ago. Businesses then were much easier to turnaround because the basics of the: ethics; financial model; business model, dialogue and valuations were relatively sound.

But now the multi-faceted tech and EDI/ESG/government funding and other "cultural ideologies" that have been overlayed post-pandemic have made businesses far tougher to turn around and become socialist or external in their outlook. In many cases, they are drifting toward victim-centric psychology.

California, the democratic ideal, is a woke culture fuelled by resources and the curse or inappropriate distribution of resources. Basically, when things are too easy, wealthy woke sets in. In California, this manifests itself as public sector employees earning twice that of the private sector.

Woke is the parasite, not the solution. It is not a scalable model elsewhere because the wealth is created by the tech giants, and as ever, the state has appalling homelessness and does not look up to its own Marxist ideals of protecting the working man.

The post-modern model is one Tower of Babel, with tyranny at the top and atomising the rest of society. It trashes all the rules and then persecutes those for standing up for the rules. Then constantly evolves ideology and language like a virus leaving senseless ideology and evil strategies unchecked.

This is the environment where the wrong people thrive and win, as well as radicalising people who don't care.

Several catastrophic ideologies are happening concurrently:

Take education too: the Welsh RSE Code now requires that all schools, cultures, subjects, and teachings must be “queered”, regardless of any objection.

The Welsh Government, led by Welsh Labour and Plaid Cymru under a co-operation agreement, has branded concerned parents “misguided” and refuses to consider there may be any basis for legitimate safeguarding concerns. This research has been carried out by citizens of Wales who feel, in the current climate, that they must remain anonymous.

We are one major disaster away from catastrophe, yet naive, inexperienced leaders follow this nonsense in a bubble. It’s comfortable because it enables the protagonists to ignore the truth. For example: gender abandonment uproots our identity consciousness and personal responsibility. It undermines the ultimate respect for an individual consciousness being steamrolled by an (invalidly) victimised group. What ends up happening is that genocide ensues because the victims rouse a “to get them before they get us” mentality.

Victim-centric psychology is the root cause of a company turnaround.

So, who is now taking the lead and thriving?

Those with the capability and enterprise outlook to drive business excellence and have the courage to make things happen traditionally are only 1-2% ish of an entire business population.

Those with the capability/aptitude/resilience to make top-end turnaround directors only probably constitute 50 or so of the total working population in the UK. Most of those turnaround directors in the TMA and or IFT lack the skills to fix the right problem because they don't see the root cause or the underlying psychology driving the strategy. Nor do they know how to change it.

Most employees, or the remaining 75%, don't care, turn up to work to earn money and progress etc. These people will normally go along with, ideally, pretty much any plan if the process is fair. But these employee types are vulnerable to following stupid strategies without much thought. Maybe their attention is more centred on pleasing a charismatic leader than what is right for the business.

Charismatic leaders become self-serving idols who distract/divert action away from the business imperatives, leaving many worshipping the leaders rather than what is right and true for the business.

At a recent talk, Sir Donald Brydon reflected that directors fall into three categories:

• looking after their own interests

• looking after the company's interests (which raises the question of what is in the company's interests?)

There is now a third category - the rise of activists with narrow imperatives that can usurp the company's interests or even the basics of the imperative to protect value and make profits.

Activism is now a serious embedded issue. Heidrick and Struggles recently reported that 93% of companies viewed EDI initiatives as more important than in 2020, and the Big 4 seem to have jumped on this gravy train and other activist bandwagons.

As I explained in my book, *The Evil of Silence*, the rise of the activist shows the power of neo- Marxist thinking in tough times: centred around emotive mantras of sex, identity, ethics, race etc.

The disenchanted might feel victimised, so find purpose in these flawed mantras and roles that have been created - head of diversity, EDI etc. Sounds OK, but the language has been redefined and centred around the neo-Marxist methods of activism, which historically have caused poverty and ultimately only led to far more violence and death c20 than any fascist ideology.

We note that the Big 4 mantras focus on EDI-type ideology and are so prevalent that their employees aren't often aware of them and therefore don't question them. This mantra mechanism creates blind spots and is a major cause of fraud. In my analysis of the psychology that caused the demise of Carillion, the overwhelming mantra of cash not profit would have made it impossible for the junior KPMG members of the audit team to speak up to the partners. So are the Big 4 purveyors of EDI perfection? I think not. They are fuelling the fire!

The clever thing about EDI/thinly veiled neo-Marxism is that these destructive, divisive initiatives and emotive mantras have radicalised those who historically went along with strategy and didn't care much. This activism disrupts and usurps normality: the sanctity of ethics, institutional structures, and valuing of the individual, exactly as the neo-Marxists intended.

But they offer NO alternative solution, and the history of such Marxist ideologies kills innovation, wealth, and ethics and creates a tyranny that eventually ends up in serious violence and bloodshed.

Violence becomes a necessity once talking stops, as it is the only way left to communicate. Social media silencing, holding people in fear, tragically immobilises the 1-2% of effective employees too.

CEOs go along with EDI-type ideology because not participating could lead to a social media attack and cancellation.

CEOs have been shown in many organisations to lack the skills to make sense of that or fend it off. Even if they do, their positions become untenable because fearful external activist investors stop investing, as was the case with the great Nick Buckley, who came under attack for challenging such ideology.

This puts HRDs in an even more heavy-handed and powerful transactional role when managing this tyranny that masquerades as rolling out values, EDI training, ethics etc. Along the way, it trashes the integrity of those who care and have integrity. Why does this ideology trash integrity?

These neo-Marxist ideologies are based upon fear, pride and lies and have no history of success; they are the opposite of Lenin, Mao, and Stalin and show poverty and violence. Hitler deployed the same victimhood, scapegoating, and terror to terrorise the Jews. A tyranny uses this mechanism, left or right, and is always based upon lies and operand, and skilled at supporting these lies until the populous pass the point of no return, at which point the people become silently or actively complicit.

So, in 2022 vs 2020, we need to understand and resolve a serious web of issues, lies and flawed actions. We need to understand and make sense of the underlying drivers. Rather than slow down, there is a panic to railroad these flawed strategies, escalating the fallout or meaning and well-being and loss of profit, which is hard to speak out against. This is the abyss I would like to explore with you.

No wonder we have a mental health crisis and skyrocketing suicide rates. These ideologues create contagions in girls and women who, by empathising with their peers, enact these gender and other ideologies out with their distressed or confused peers. In this way, actions force people's beliefs to align, which causes a real psychological mess, confusion at a fundamental identity level and stress.

Where are CEOs at?

In 2022, we offered an opportunity for the 6000 executives in our network to join a cohort to learn and lead the turnaround of the UK.

We learned:

• these CEOs weren't coping with their own schedule, so could not cope with taking an enterprise view.

• they were too fearful to learn for fear of exposing weakness, even though the forum and agenda are exciting, meaningful, and great fun.

• some CEOs were interested in the vanity aspects of joining an elite group.

• others were looking for external parties to take ownership

• and some CEOs did not want to take ownership of their own actions.

That said, we found a handful of fantastic and courageous leaders from the 6000 or so who form part of this venerable network.

So, what's the scenario now, and what does the problem look like?

What do you believe in or care about?

Even Putin recognised that this woke EDI parasitic thinking is the same neo-Marxist mechanism that wrecked the ethics and crux of the USSR. He is right to challenge our ethics on this front.

I am sharing this to try to articulate why this ideology is a problem, and in the context, we have strategies to get you back on course.

The solution is fun and rewarding by getting you back to humanity.

Why don't most of your staff care?

The issue with people who don't care is that they don't find any substantial meaning in what you are proposing to do, so when things get tough, they have no resilience, and they become pointless.

So how can you create meaning in the current environment and make people care? You can value them for the small things - relationships, open dialogue, and different views.

Yet with technology we are at a tipping point of the human – human empathy and connections that hold us together, keep us on track and give us resilience when things get tough.

Our sanity is not defined in isolation. It is defined and supported by those we have productive relationships with around us. Yet, with the deluge of initiatives/fear, we are losing the ability to make sense of the issues and get a grip.

The relationship justifies profit and pricing too. No customer intimacy and relationship = no justifiable profit.

How many conversations are we having with beneficial adversaries, or have you fallen out with them?

What do you put your faith in?

Jesus said the proud are an abomination. I think many tech businesses are based upon an ultimate form of hubris and pride, and if it isn't working, rather than accept that it's creating more problems than tech solves, the self-serving mantras continue: "More tech will fix this", postmodern undermines basic truth, "male/female is a human construct", then the ultimate madness, "AI will fix bias etc," when AI operates purely upon bias, and there is no ethical basis for AI, nor are we programmed to be able to do so. Ethics are highly nuanced and complex, well beyond our understanding for sure, based upon a higher power well beyond our capability.

Employees aren't stupid, so what is the experience of the abomination that makes up pride? An abomination in a business causes loathing and disgust and a feeling of hatred. When you confuse this with the hubris of defining your own gender without challenge, you create a fragile infantile scenario whereby you will inflate until you burst!

Pride prevents us from seeing our mistakes, so we are stuck. You can see this in the shifts in mindset during the pandemic: people realised the value of relationships, valued what they had, valued slowing down and the link with sustainability etc.

Yet since the end of the Covid pandemic, we have gone back to where they were, in fact, worse, ignoring such learning as the crux of society is relationships, reverting to the mass illusion of the self-centred material values of before, growth and domination, stuck in self-centred pride, as well as excessive consumption. Excessive consumption is, of course, the root cause of trashing the planet.

Take plastics consumption:

If you think you know better and strength is based on your self-reliance, see how you get on.

I think we are witnessing the symptoms of this hubris now. In even mildly difficult scenarios, self-reliance is not a robust strategy. When things get tough, survival depends upon looking out for others, not you as a centre of wisdom. It is love or acting in others' interests that holds a business together.

People in tough times need a robust purpose to keep going, yet these EDI initiatives reduce everyone's identity and sense of self. This is what we derive our basic strength purpose, meaning and strength from. Tragically EDI is replacing the very simplest pillars of identity with a set of false beliefs, group thinking, accusations of white privilege, systemic racism etc and anti-bias training for those who don't fall into line with these untruths. This tyranny is quietly silencing people, killing the basis of our humanity and reducing you to someone who doesn't care. The post-modern system doesn't care or value you as an individual either, its purpose is to confine you within its lies.

Brand is about relationship, experience and meaning, and meaning is as much a spiritual issue as anything else.

Whether the environment is ordered or chaotic it makes no difference. If you have lost any engaging meaning, you are stuck, people won't care and/or the wrong people thrive.

Your spirituality, legacy, and ethical structure define a framework you work within – seen and unseen. Unseen because it's complex and too simple and familiar to rationalise, such as humility, gratitude, appreciation, marriage, your sex, institutions, grace, and love. When these pillars are attacked, they are difficult to defend because they are so familiar or complex, or like Jesus's teaching, "love your enemy" beyond your experience or relevance envelope.

Humility and gratitude are being overstepped by EDI, ESG, WEF ideology. This ideology has excessive pride and ignorance of the virtues of these legacy pillars, evicting faith. Evicting faith is reckless, or pride in drugs, because no civilised society has ever prevailed faithless. All faithless societies implode.

Fear is the tool commonly deployed by ESG, EDI, big tech and the WEF because fear holds people in silence and supplies the opportunity to manipulate people to suit an ideological agenda. Fear accelerates the progress and evolution of these ideological fear-based systems. Endless iterations of the ideology without reason or challenge create hysteria which holds people in fear of speaking out.

During this process, what is happening to employee well-being and purpose? The perpetrators of these ideologies rush to conclude their strategies because they realise, they are not smart enough to create the ideology. Their ideology is flawed now. Can they resolve the issues or hold the fallout?

COP27: fear and no communication with China

– the stasis and farce

China and India are not attending COP27. China has formally requested to drop out of the COP targets up to 2030, to provide them with a window to trash the planet and our Western industries by 2030 by accelerating their creation of C02. Then once they are in the driving seat, reduce emissions to 2050 by which time we in the West will have fewer key technologies and industries. The West accept this abstinence, so they should impose a tariff on Chinese products to balance the manufacture to the West and to create a level of the #zerocarbon playing field. Our industry playing by the Marquis of Queensbury rules makes us far less competitive in what is a cost-sensitive market!

This is a good example of how a fear of China prevents tough conversations and keeps dialogue happening by letting them opt out. It makes a complete sham of the 1.5C target, above preindustrial levels, experts note +1.5C has almost been reached already!

The UK's zero-carbon strategy will never happen because much of the technology is flawed and the implementation structure relying on local government execution is ridiculous. The operating model needs to be changed. The local council chiefs are not capable of selecting the technology, creating the financial models, and so woke that there isn't much chance of re-aligning this strategy either!

To make the situation worse, WEF types are proposing to give up and base strategies upon unproven carbon extraction machines to take carbon from the atmosphere. This is crazy-scale ridiculous, and technologically unproven. We need to get back to basics and the root cause of the excessive consumption shift, which is our own making – China.

The woke beartrap is why there are so many issues.

Because corrupt ideological solutions are being deployed, the wrong people are thriving, so we are creating mass rework. This is manifesting itself in a disregard for dissenting others' views that would normally re-align strategy, causing increased stress, mental health issues and escalating suicide in men/eating disorders in women.

The cancelling labels lies, such as falsely being accused of being racist, anti-trans etc, or the all-encompassing mantra of taking offence creates cry wolf, where the real issues are being covered up: child abuse, real racism, flawed technology and an unwillingness to challenge political structures.

Without leaders who value and relentlessly search out the truth and have the courage and support to make tough decisions, we are heading to hell. The post-modern model is designed to infiltrate the top-level structure and then create division between friends downstream. In this way, the UK has replaced high-integrity leadership that cared with a woke, infiltrated structure, which does everything to obfuscate the facts and any rework they create, thereby normalising their challenges.

This protects their fiefdoms by enabling them to ignore issues and cancel those that don't fit their self-defined ideological pillars or dialogue. These people in the public sector have little or no technical, financial, operational or leadership ability.

Recruiters are forced to recruit to fit the public sector culture - anti-bias characteristics to fit the neo-Marxist outlook - rather than recruit people suitable to achieve a successful outcome.

This environment is untenable for good people who value the truth, and those who stand up to it will get at once marginalised, labelled, and side-lined.

Processes should not be organised for the exceptions, such as making the trans population mainstream, nor do trans people often want this treatment. Exceptions should be handled by exception, or else gender dysphoria as a mental illness and a cause of mental illness in girls will spread by them empathising with their peers.

So, creating a process that forces everyone via this trans or exceptional route creates nothing but division and keeps control of the neo-Marxists who have infiltrated organisations. It does not follow that if you don't approve of making this ideology mainstream, you are anti-trans etc.

Respecting institutions and the pillars of society

Various pillars in society make society work; government, marriage, identity, male and female, race colour blindness, grace and love, humility to listen, valuing individuals and different views, taking an interest in others, and maintaining dialogue with enemies, even when faced with our own difficulties. These pillars are so familiar that we hardly see them, and institutions have taken 100s of years to evolve to get as far as we have, so throwing them away without another solution is simply reckless.

Mature societies that have it too easy tend to become self-righteous and proud and so turn in on themselves because in having it too easy, they tend to make up their own rules. We are not smart enough to make up our own rules, so we get into a mess or become catalysts to tyrannies.

Ideological rules become tyrannical and create blind spots and a lack of grace, trashing humanity and the basis of our individuality. The ability to value and enable our own beautiful, unique contributions is the art of leadership and management.

We naturally see the world through the lenses of us vs them; it's a basic intuitive survival tool and part of our humanity. Yet these encouraged woke pressures to conform to the ideology cause people not to be innovative.

Grace and love matter because humility suggests we don't know any better, so we should be asking what can I learn from you?

The cost of this utopian thinking leaves the thinking unchallenged, reinforcing our natural snobbery – fuelling the self-fulfilling prophecy and grouping that suits fragile echo chambers.

The labels of utopia also create an illusion of bigotry, for example, pathologizing men as being toxic.

The same tools that protect echo chambers are critical to protecting this tribal group thinking.

They are also convenient and useful tools to foster in social media and echo chambers fuel the very same divisions in crypto, WEF-type globalised thinking, DeFi, etc.

Social constructivism denies social imperatives such as a wish to improve ourselves, our male and female biological features, valuing things vs people and cultural differences. Our biology is

supported by cultural norms such as women being less sexually promiscuous, and people's means of improving themselves are different and not just driven by power.

Societies of leisure and plenty are where eating disorders are experienced by women; anorexia only appears in societies of plenty. These illnesses only arise when people are NOT worried by what matters.

So, it is often immigrants who fight for what is right, leaving the cosy that haven't seen hell, blindly following, and creating untruths.

It doesn't take many radical people to bring down society. But people do self-sensor, and there should be no forbidden statements, yet people are afraid to voice opinions.

The silent majority must speak up to marginalise this cancerous ideology, yet people lack the courage to follow their convictions. Winning thinking creates relationship, empathy, and the realisation that people are different. By taking NEW action, the candidates experience humility for themselves and a comforting realisation of their interdependence. This requires courage to step over the edge and leadership to execute and help.

Mantras block critical thinking.

It is a gift from God to appreciate what you have, even at the simplest level, water coming from the tap, putting the light on etc. Yet for the last 40-50 years, the mantra of **"more"** has become prevalent, driving a focus and curse on growth. This growth has been fuelled by cheap products, low inflation, and debt. China too has kept prices low, intentionally trashing core industries such as steel and manufacturing. Looking at the pricing of a bicycle over the last 20 years, the price has remained static, so broadly, prices over the last 30 years are supressed by China by 220%. So far, more inflation is inevitable.

With this growth, we are now paying for this stupidity with damage to the planet and excessive consumption – as previously pointed out in the global production of plastics.

Growth and domination too have no ethical roots, so when combined with the EDI and ESG/ WEF/ideology and the relentless manipulation by technology, businesses are fuelling a mass loss and crisis of meaning and purpose.

Yet these pervasive mantras continue, as Liz Truss's debt-fuelled uncosted growth plan proved, the UK no longer has the patience from the markets to carry on borrowing recklessly to return to growth. There was no evidence that these tax cuts would lead to that growth either.

The reality is that the UK will get much poorer because the growth delusion or bubble has burst, and the tough financial issues need to be faced by the UK population, appearing in increased taxes and inflation. Yet the pointless technology businesses crypto, web3, DeFi, etc continue to pillage the productive parts of the economy in pointless ways. But these digital SAAS businesses don't require working capital to grow, so are relatively inflation- when scaling. This fuels investment in their direction and away from productive businesses, which are far more difficult financially and operationally to run in an inflationary environment.

The underlying assumptions around the UK Government's zero-carbon strategy are centred on reinvesting savings. This is unlikely, as well as having a diminishing return. The technology being deployed, hydrogen, EVs, when you look critically at the carbon impact end-to-end of Evs, for example, are making the scenario worse. The technology and definitions of the technology applications, for example, colours of hydrogen, are misleading - the psychology of the claims producing “green” hydrogen is misleading because of the opportunity cost to use that “green” energy to replace the hydrocarbon element. In terms of the applications, the fundamental energy cost to liquefy hydrogen to -270C, almost absolute zero -273 makes a nonsense of using hydrogen in planes. So zero-carbon and zero emissions are two different things and must be considered – end-to- end.

Decarbonising methane produces hydrogen, but that presents issues with pipework because hydrogen is a tiny molecule, basically a proton, so it leaks and doesn’t burn. It explodes due to hydrogen’s fast burn rate. That is why hydrogen is used to fire rocket engines.

Critically thinking through the options makes even the claims or purpose of “green businesses” dubious at best. Very clever, innovative end-to-end strategies to use all the offtake in “industrial ecosystems”, such as graphene and hydrogen, to reduce energy consumption and/or materials and use the graphene to reduce the use of plastics or concrete.

The main opportunities and priorities are, in fact:

• growing and looking after natural systems, we know very little about these except we are

losing them, forests, peat bogs, soil etc

• notice the hubris thinking we know better!

• reducing consumption (this is a psychological issue driven by the growth and mantras of more hard-wired into economies)

• notice how this challenges the mantra of “growth”

• use technology, largely reducing consumption

• generate energy in a zero-carbon way eg, nuclear and or switch from coal – gas (50% less C02

with gas vs coal power generation)

• the specific issues need to be managed and not overwhelmed by the negative generalisations:

related nuclear safety, fracking fears.

The big issue here is how undeveloped economies will grow and how we can set the standard for them to follow because right now, we are exactly the capitalist growth/greed model that they should AVOID. Yet, fossil fuels do play a strategic role in a practical economic solution.

So many of the current mantras justifying tech strategies are creating industries that don't make sense, so they compromise meaning and purpose and their ability to attract the best people, let alone achieve zero-carbon.

The circumstances and or the difficulty of circumstances are immaterial. The integrity of the decision makers and strategy, fundamental honesty and the heart of the business make the difference.

The breadth of what now defines a "good meaning" is achieved across a new set of dimensions post-pandemic, see top 7 later. Yet the mantras of growth and domination remain overwhelming, exasperating people, meaning and good strategy.

Most of the motivation behind growth is improving happiness. Happiness is centred upon happenings – largely through acquiring assets.

Joy, however, is an inner-biblical which is based upon grace and love. This is regardless of circumstances or happenings. Businesses that lack grace, love and or meaning or sound purpose undermine the integrity of people who care and have a good moral compass.

Happiness is related to pride/hubris because it assumes we can take charge of controlling our environment in a self-righteous, all-knowing type of way as a means of regulating our happiness. So the happiness obsession gets us into a mess.

To the contrary, the experience of pride places a burden upon us that we can neither live up to nor cope with. Worse, it assumes that we know better and don't learn from mistakes.

Jesus said the proud are an abomination. An abomination is:

• a thing that causes disgust or loathing

and/or

• a feeling of hatred

Hatred and loathing are two key aspects of the intolerant, proud, woke culture. Woke is driven by pride and redefining truths. Or worse, postmodernists assume that the only truth is that there is no truth, so truth, and therefore reality, is a subject man-made construct. This creates the perfect platform to fuel and justify untruths and division around gender, racism, etc, without a rational check or balance.

During the pandemic, there was a heightened experience of humility and love between people, post-pandemic sadly, the shift has reverted to growth and domination, with an increased sense of desperation and strategic ethical misalignment. For example, the German Government, having fallen foul of Russia and paid the price, carry on regardless: in September, the Germans sold a share of a Hamburg port to the Chinese. Yet China is a far greater threat to a meaningful and worthwhile purpose, so contrary to learning, Germany is continuing to fuel tyrannical regimes.

This inability to change, in part driven by growth and domination, is like an alcoholic desperately reverting to alcohol with increasing desperation. This pride creates hell.We are not designed by, nor do we know better than, God. Surely knowing better than God is an absurd notion and the basis of an absolute abomination.

So, where therefore do we find our ethics? How can we rationalise what is right or wrong? The fundamentals are driven by the first commandment: “love God and love other people.” It is a heart thing, so if our hearts are in the right place, we can ratify what is right or wrong. Or else how do you ratify ethics’ purpose or meaning, or make this up yourself?

Or if you don’t love God, and if we are love machines, the question is, what do you love?

Why relationships? Jesus is all about relationships. He died on the cross to take our sins and reconcile us with him to have a relationship with us.

Profit is important too, so we don't become indebted or tied to some evil or nefarious influence or nation with conflicting underlying interests, such as Russia or China. Yet we are told to love our enemies. Love fixes things.

So strategic stupidity, laziness, and hubris are causing us to engage with nefarious forces, so relationships are getting trashed, and wars and blackmail will rise.

The silencing and wokeness of calling out nations cause a cry wolf to cover up for maybe being accused of racist challenging by China (CCP). The complexity of the strategic drivers is hard to follow but the underlying traits causing strategic distortions are similar, distorted dialogue, language, ideologies, truth, free speech, response to counterviews and attacks, and redefined meanings etc.

Ethics vs Narcissism

Ethics are the moral principles that govern a person's behaviour, or the conducting of an activity based upon well-founded standards of right or wrong that prescribe what humans set out to do in terms of rights, obligations, benefits to society, fairness, or specific virtues.

So, who is qualified to set ethics?

Narcissism, contrarily, is extreme self-involvement to the degree that it makes the person ignore the needs of those around them.

So, how can we get ethics right without creating an ideology, self-reliance, or tyranny? Christianity shares a proven model; freedom and counsel from Jesus on a "come as you are" basis, from which ethics follow.

Christianity is a faith, not a religion or a rigid set of rules to follow. Christianity requires you to let go and then rely on Christ. This frees you from your sins, and the ultimate relationship is with Jesus moving forwards. The 1st commandment is to love God, love other people, and the rest is a heart thing that follows.

Alternative religions are religious and ideology centric, so they become a tyranny. The structure of the ideology, whether it be woke, Hitler, Stalin, or Mao, is common:

- victimhood

- scapegoat those that don't fit

- terrorise – and or cancel

So, who thinks they know better than Jesus?

Tech creates a monster in people's heads by having a mantra-fuelled cult-like following, "more tech will sort it out", "DeFi", "decentralised utopia, redefining countries etc", "social media connecting", safety "vetting of acceptable speech – silencing and avoiding conflicting ideas" all lead to an increasing reliance on AI, to "sort out bias" (when

AI is run on bias!), and social media is designed to manipulate and addict by triggering dopamine, comparison or disappointment. The

addictive aspect is based upon the anticipation of, not the receipt of the item, trapping people in an unfulfilled zone of anticipation and atomising.

The most damaging rewards are those that need no effort and/or are unearned, such as virtue signalling and inhuman cancel attacks for those that don't fit the ideology. These are hardly congruent strategies that fit with inclusivity or "EDI."

As a lot of contact is via tech, the language ethics and dialogue are regulated increasingly by AI/tech. So, their self-defined ethics conveniently silence those who disagree with their broader, underlying ideology of social media control, and tyranny of tech/crypto, WEF etc.

Defining and redefining the model is the ultimate madness too. The problem is too complex and irrational, and it changes every nanosecond. So, the solution lives with us taking ownership of ourselves and our ethics, sanity, or strategic steer, depending on good productive relationships with others.

Woke ideology started off well by trying to create a means to rationalise what was right or wrong from several options, but evolved to disregard the inconvenience of individual views, and the underlying motivation and woke derives its power and hold based upon creating division centred around emotive subjects such as race, gender, sexuality etc. It starts by infiltrating structures top down, then takes hold of the dialogue and creates division down the ranks to silence, control language and gain power.

"Over time, as governments began to separate themselves from religion, many responsibilities of the church became transferred to the state. The gradual progression of socialist and Marxist thought, meanwhile, further increased the divide, while at the same time expanding the remit of governmental reach into people's daily lives.

Taking into consideration the challenge of rampant illegal immigration, the divisive nature of "woke" ideology that pits citizens against one another, the criminalisation of speech that constitutes "hate crimes" law, the softly-softly approach to fundamentalist terrorism, as well "transgender" infiltration into women's sports, locker rooms and restrooms, media race-baiting and so on, it seems we have a bit of a problem on our hands."

Quote from "Socialised medicine is no cure" 2022 Gatestone Institute.

When things get tough you need a robust meaning and purpose. An empathy with others is part of the love that holds teams together, or else the wheels fall off.

Now more than ever we should consider spirituality because our spirituality drives our emotional well-being and is the crux of ethics. Ethics are a central pillar to strategy. A business in a distressed scenario exposes the nonsense or that activity to stop. By becoming aware and being able to read the language and nature of the dialogue, you can achieve an indicator of the drivers that got the business to where it is or caused its success – and or what is the core business.

Treating the symptoms of a business failure during a turnaround, failed leadership, wrong actions etc, often leads to the wrong people being fired. People are often just not in the right roles, or maybe their position had become untenable due to the mantras and blind spots. It is very common that issues cannot be discussed or shadow falsehoods. These falsehoods become so familiar that they had become part of the furniture unchallenged.

These blind spots, falsehoods or ideologies destroy a business purpose and thereby trash the motivation to stick to the party line when things get tough. It just doesn't work because you can't scale the existing culture; people just go mad.

So, without a CEO with a sound baseline of truth and ethics devoid and/or aware of any ideology,

there is no means of preserving a meaningful purpose. The corruption is that the woke ideology creates a perfect corruption in its attempt to redefine "the truth," based upon its fragile aggressive moral high ground, emotive topics, and tyrannical principles perfect for creating blind spots.

A tyranny is a real issue because evil is supported in silence. This silence corrupts good strategy because the ship is left without a full rudder, the ability to face reality, and our overall sanity as a business team is supported by others and our peers.

Tech creates a parallel and futile universe. The critical factor in deciding viability is profit or people's willingness to pay. Profit is irrelevant with the growth and domination focus on a tech start-up due to the excessive capital injection. So, celebration and company valuation are defined by how much money is raised as a % of the equity, NOT profit made. This measure of value creates insane valuation multiples of 100s of EBITDA completely defined by cash raised for a percentage of equity. A loss-making business of this type is a Ponzi Scheme, so it just collapses as and when it runs out of cash – so the focus is on Series A-D fundraisings!

A peer and fan of the "rise and fall of cultures" recently suggested that there was a 1 in 6 chance of Putin using nuclear weapons. In a world of such utter nonsense, one wrong move could change that to 1/1 without rationality. The human and business challenges are not some sort of biomechanical challenge either. People are NOT machines who call for an endlessly complex tech-type solution. The solution for us is far simpler than that, decided by our ability to keep relationships and productive dialogue regardless of circumstances.

It's important to value the truth and face facts rather than distort and delete them. Because the truth includes errors, this requires forgiveness and sacrifice to reconcile our errors of the past. Without forgiveness, the alternative is anger. The redefined frameworks of woke language and causing zero offence create a toxic mixture and seedbed of complete collapse we are at risk of in the UK corporate and public sector currently.

The rework coming from these flawed, tyrannical tech strategies and weird ideology are misguiding strategies, which is great for the Big 4 as it creates endless work and perfect gravy trains.

The ultimate madness is to think that we can understand and control the emerging complexity and woke mind viruses appearing on all fronts fuelled by the powerful rich, WEF, big tech owners etc. It's time to stop most of this. It doesn't fit with profit and relationships.

To illustrate the need for elementary critical thinking, let's take rocketing energy pricing. Energy pricing is driven by the marginal pricing policy, so rather than ring-fence the excessive provider cost, the rest of the energy producers must increase their prices to follow suit. So, the whole industry is making a fortune in a scenario where our imports in the UK are negligible. This is easy to sort once you find the root cause, yet the government subsidises the excessive cost rather than addressing the root cause of the marginal pricing strategy.

The good news is that tough times expose nonsensical business models, so they do get found out in a crisis. Big tech business shareholder returns are falling year on year, nor are they creating new industries to replace this profit gap. Rather they introduce feudalism and their economic rent on the productive players such as: Apple, which imposes 15% fees on service providers using its dominant app platform, and Google, which controls SME search marketing and sales. Whilst they have vast scale because these tech businesses are easy to replicate, new entrants (such as Revolut or Monzo for example in the banking sector), can come in and chop pricing easily or give away services free of charge to gain market share, creating an overfunded, growth and domination centric race to the bottom, just as they did.

Keeping up and the tech catch-22!

People's ability to process this overwhelming complexity in a tech-driven environment, save intuitively knowing what right and wrong and taking responsibility for their own actions is, is limited. It is only via local and global collaboration that we hold these systems together. If we don't act, who exactly will instead?

Communication, through valuing and taking an interest in different views and relationships is critical. We also need a robust ethical compass to check our actions against, not self-made values!

Jesus died on the cross to have that intimate relationship with us and help us navigate these scenarios.

Jesus is all about freedom. It is only when people are free and given an opportunity with accountability that they begin to do the right things. Germany, on the contrary, is commercially wedded to China and digging deeper so it seems not to have learned from the cost of dealing with a tyrannical regime from the debacle with their reliance on Russian gas! It requires courage and humility to undo things.

The WEF trajectory is heading to simpler industries, more automation, less opportunity, more central wealth, less local income, hubris focussed and a narcissistic strategy without a robust purpose – a race to the bottom.

The cruellest factor is that IQ is fixed. So, with the WEF strategy creating more exclusive higher IQ-dependent opportunities, increasingly, people are still excluded from the job market, isolated from society and unable to contribute.

The catch-22 occurs when the speed and sophistication of parallel AI systems and the increasing fight between the big tech companies keep us out of the picture. They will continue to control dialogue, ethics etc, and we simply won't be able to keep up. We are at the early stages of this dystopia now. It is subverting our relationships and own intelligence, reducing our human intelligence or ability to fix problems.

The rise of the strong/overwhelming problems

History suggests that strong individuals like Aleksandr Solzhenitsyn resist these broken systems and bring the truth into the light for others to catch on to. Aleksandr, in this way, collapsed the Gulags. One hopes that the same will happen to the Chinese Communist Party, Iranian or Russian type tyrannies, or the killers of Mahsa Amini in Iran, which sparked populist uprisings in Iran; although the Iranian Government has just 15000 executions of freedom fighters to terrorise their population.

It is difficult to know how Solzhenitsyn types could get traction in the current social media environment. Many of the best people are unpopular vs what gets traction and are far more likely to be silenced or their outspoken speaker trashed - Trump (anti-democratic), Johnson (liar, party gate etc), Suella (inflammatory). The labels are deadly, and key woke tools to trash reputations en masse, rather than give these brave people grace and love and support along the way.

Because of the increased inability to communicate and have tough discussions, legislation is appearing to control offensive free speech!

Promoting woke ideology creates fear on many levels and cry wolf around real racist or sexist issues. The fear of offending creates bizarre outcomes, mental health crisis on the one hand and silencing on the other, thereby preventing people from acting in alignment with their beliefs, inducing real stress.

So woke ideology is the ideal platform to obfuscate the truth and dialogue, leaving no chance of having tough discussions, and the good people who care to get angry and act badly, inflamed due to endless attacks by the media such as Boris and Suella etc.

To make sense of these scenarios, we need to promote free thinking – to break this deadlock alongside forgiveness, grace and love.

Silencing good people in this way will drive them mad and liberate the idiots. We need to liberate the good people who value the truth, or we will continue to see evil take root and the wrong people thrive unchecked.

These misaligned strategies, fuelled by woke, create a lot of reworks, noise and coverup, a bit like using lying as a strategy - it's too complicated to maintain, and liars lose their thread. Therefore, avoiding dialogue or subverting integrity or the truth is a critical tool to maintain the lie!

Woke, or post-modernism, is professional lying, the perfect tool to create menticide.

The cost of not having productive dialogue is high: out-of-control tech strategy, crypto taking cash from productive economies, loss of sight of systemic trends eg, employment, the balance of payment issues in the UK, China trashing the EU industry, an escalating mental illness crisis etc.

It is much harder to make money in a productive economy. It's not fuelled by excessive debt or misleading mantras. It must meet a real need, sell at a profit and generate cash positively or else, unlike an overfunded tech business, the productive business dies.

The productive economy requires 100% teamwork to avoid screwing it up. People must turn up to work on time. The company needs a robust manufacturing process, collaboration, and direct feedback must be rooted. And the business needs to live within its means or go bust.

To make money and sell products requires real rigour, which means relentlessly searching out the facts and fixing problems at once. This is far harder than raising and burning investors' money to create some type of network effect to get the wheel spinning. The wheel can also spin the other way.

In fact, the logic of a tech valuation is that the faster you burn the cash, the higher the valuation, which is just mad.

Successive attempts such as Revolut to attack the retail banking market will lead to a gradual destruction of value. New entrants will just replicate and not charge to get the wheel spinning and trash the incumbents' profits and value. These businesses do not engender customer relationships either. No relationship – no profit.

Tech is all about control of attention which requires addiction or hijacking of the dopamine (anticipation) or serotonin (one-upmanship/comparison) influencer system. Dopamine is triggered in the same way as heroin and alcohol, so social media leads to addiction. Social media companies have tried to change their algorithms to a more human-friendly basis, but the attention on their platform stops.

Once the tech addicts us, people get stuck in their respective echo chambers as designated and orchestrated by the tech. So there is no challenge to the thinking or critical thinking, just protection of interests and protection from attack within the bubble, fuelling the cash-burning madness.

Investors get corrupted with the focus on customer acquisition and network effects rather than profit and relationships, thus usurping their interests in financial rigour and gross margin.

To support these businesses, the echo chamber or bubble must create its own weird dialogue to avoid reality and subvert, without checks or balances and certainly no critical thinking. These businesses' products, such as crypto, hardly hit the ground with the mass psychosis/hysteria being fuelled by false mantras such as "tech is the future", "tech will resolve" (C02 capture), "gilts have fallen", "governments have been stealing from you by driving inflation" (crypto).

What is critical to confirm these business models is to develop the skills to find these mantras because they are driving the underlying thinking and strategy. In that way, you know what to challenge and how to challenge these scenarios ie, the problem to fix.

Propaganda is another description of such language. China is the propaganda expert, chipping away at the credibility of Western institutions and infiltrating established institutions such as Confucius institutes in UK industries.

China's population demise is a real threat to China's success. This drives their expansionist strategy to dominate Western industry in the country, for example: trying to buy Aston Martin, owning Lotus, owning MG cars, China replacing UK's Mini manufacture etc, yet their own citizens don't put their trust in investing in Chinese companies, or Chinese property now. Tech and the Chinese culture are like revenge for the opium wars imposed on China by the West. China's only means to grow is now via controlling other countries and their industries' infrastructure. This is all part of China's five-stage strategy to 2050. Whilst the Western democracies muddle on, China CCP has a plan to control and attack our business and democratic interests and the autocracy to remove dissent or speed up decision-making!

Standing for nothing and tolerating anything is hell and stressful. That is the paradox of liberalism.

Love is about willing the good of another, so not tolerating anything, and protecting victims of criminal acts, or a society without laws will collapse.

It's important to be inclusive with people, not necessarily their ideas. Love means you want good for them, not hate. Love thy enemies. There should be no obligation to tolerate another person's actions or beliefs.

So, by separating the person from their ideas, you can discuss their ideas freely without labelling them, without violence, exclusive perspectives, just critical engagement, and public argument.

Argument is middle ground and love enables you to engage with people's ideas with humility and interest, lovingly not violently.

Loving everybody does not mean accepting all their ideas, be or do what they want, that is not necessarily what is good for you or them.

Faced with a disagreement, you can separate a person's ideas from the person, so it's not a personal attack or challenge. The idea you hold may be daft, but it doesn't follow you are a dumb person. So, separate a person from thinking and hypersensitivity. Most of the ideas that come into our heads are wrong.

The difference between attack the idea, not the person, is important. It's the difference between assertion and aggression. Attacking the person denies their identity. Identity is lost in flawed language and ideology.

DEI should foster a healthy sense of intolerance with boundaries kept; this is what I am for, society is set on bad values, if everything is tolerated, it stands for nothing, and society collapses.

Abiding by a strict code is attractive, but football is impossible to play, it collapses without rules.

Christian faith should lead to freedom.

We have had an extended period of excessive comfort for the last 40 years. A period of ease where people have been able to exist without faith, easy times allow people to get away with reinventing their own truths. Untruths get found out and implode, so there is a curse and dangerous hubris in comfort.

Socrates, Plato and others all struggled with this challenge, "what are we doing here?". They all noted the importance of local and global collaboration to survive.

So, who has the grace to make tough discussions when their self-perpetuated circumstances are overwhelming them?

Jesus was by far the most influential leader to ever have walked on the planet – he set the date, 2022. Jesus answered this key question, "what are we doing here?" and the destructive force of pride: self-reliance and self-righteousness that destroys society. Jesus said the first commandment, simply love Jesus and love others. There is no more important commandment. That is a simple purpose and courageous journey because it throws self-reliance down the loo.

Jesus gave his life to provide the opportunity for people to have a relationship with him. This is not some sort of autocracy because the handle is on our side of the door, and he loves us the way we are.

Fear and intolerance are the woke mantra, which is the antithesis of Jesus's teaching of good, beauty and truth. Valuing truth took a long time to set up, and fraud in tough times was the driving force behind the emergence of the Co-op in the 1800s in the UK. The 1800s was a tough time when fraud, such as putting chalk in flour to bulk it out, was common practice.

The Co-op was a UK business rooted in Christian ethics, local support, and collaboration to get back to truth and use business as a force for good and a means of rebuilding the community. The Co-op brand still thrives today.

Mantras – getting stuck in the bubble.

The steam engine technically would have been better than an ICE vehicle, yet the dominant psychology driving the decision-making that selected ICE technology over steam, because steam was associated with heavy locomotives and internal combustion engines, were small and nimble.

The way we handle challenges that we can't cope with is to normalise, using generalised language to conveniently justify our status quo, undermining the fabric of society, democracy, dialogue, ethics, and communication.

For example, technology environments are driven by the growth and domination psychology, which dupes investors, causing them not to check gross margins when profit is the rationale and source of a business's sanity.

There is a spiritual dimension to this, diabolical means demonic. The devil is described in the Bible as the deceiver. So, this loss of truth and loss of value of the truth is mainly a spiritual issue to resolve. Yet woke seeks to trash faith and spirituality by dismissing them and creating fear. Biblically, fear is a sign of the devil, and Jesus is love.

So, in this period of self-made mantras and self-generated rework, are we really interested in what others think? Can we be bothered to have tough conversations? Without these checks, how can we move forward to find what to STOP doing to root out stupid strategies, which could appear to be destroying value and creating a race to the bottom in terms of well-being, profit, and relationships?

People are very bad at undoing things because good/bad habits become permanent habits after 66 days, and pride makes it very difficult to accept that we may be wrong.

History suggests that the cost of the c20 ideology of Marx, Stalin, and Mao was the killing of 120m people because once the dialogue falls apart, the only way to keep the ideology is by labelling people as evil or through violence and silencing.

Not valuing the individual is at the crux of neo-Marxism, of which the mechanism is victim, scapegoat, and terror. The Nazis used the same method, blaming and scapegoating Jews, Marxists and the privileged. Maintaining these ideologies requires a tyranny defined by protecting the flawed ideology.

The computing industry collapsed in the 1960s in the UK because women were not allowed to take leadership roles in the industry, and it was illegal to be gay in the UK up to 1967.

When I started my career in the automotive industry, the purpose was clear and meritocratic - make money, compete and be world-class. This was at a time when China produced junk, so it was clear outsourcing various low-value aspects to China made sense.

Manufacturing became unpopular, and difficult to defend its existence in the UK, despite the obvious economic advantages of local UK manufacturing. The rest of the UK economy grew on the back of a service economy and financial services. These are luxury industries on the back of the seedbed of manufacturing, but alone, service economy and financial services really deliver nothing much to defend.

Intermediator models the trap.

Take EVs vs ICE vehicles. EVs are higher cost (and unaffordable for many), simplified supply chains, with a common charging infrastructure, and ultimate remote control, which tends towards a financed, owned, and vertically integrated intermediator model. The product car could be owned by the intermediator and rented to the customer on a pay-as-you-use type basis.

The same applies to booking.com in terms of renting property, which takes 15% or so of the margin.

Because these intermediator business models control the sales channel, they impose a feudal system and economic rent on the industry, alongside vast profits with little work/value or risk.

China runs a lot of these business models ranging from bikes to taxis, some of which require vast working capital. China is less focused on ROCE.

This central control of these intermediator models and their converging technology and AI tends to dominate. It's a tough trap to escape because the intermediator controls sales and network and scale. Google, for example, design the cars too, and so these Google vehicles are specifically designed for their own business models.

This intermediator model lacks differentiation but centres on control of key assets, so the consumers are trapped, unable to buy their own car because they are too expensive to buy vs ICE vehicles. Further regulating comes from controlling the supply chain, for example, the Li-Ion raw materials, so the intermediator model plays straight into China's hands as a supply solution. These models historically have very little indigenous local benefit even when Chinese made and designed cars local to the market eg MG in the UK etc.

These tech intermediator industries have no relationship, or lasting differentiation, so will compete on price and create a race to the bottom. This is an attack model which suits China well, alongside controlling attention online to control the sales channel too.

The increasing simplification and frankly boring nature of these tech companies make it very difficult to compete because the vehicle is trending to becoming increasingly undifferentiated, so little price/brand value or even consumer relationship. Very hard to make money too. VW historically have made losses on their EVs so far. China can stand these losses, so they are a key tool to attack Western businesses that will iteratively exit these markets because they run on China's terms, control, and domination.

The support networks of EVs servicing garages are greatly simplified too, vastly reducing the work created. So, if the work completed in the UK automotive industry is controlled iteratively and increasingly by China, we are in a very bad place to retain local government control because we will have increasingly fewer profits to provide public services. Therefore, increasingly, governments have become a laughingstock, clearing the way for China and crypto (anti-inflationary) type madness too.

Governments do play an important role for citizens to fight national battles on our behalf; they give citizens clout. So, the notion that DeFi-type industries help democracy is nonsense because the central government support will be usurped in their utopian blockchain/web3 models, creating atomisation of the individual, alongside rated voting on everything. This will destroy national identity and solidarity, leaving national governments vulnerable to completely losing control and fewer central services to look after their poor or disadvantaged people. A democracy is a serially voted party. So, a serial autocracy for that approved period needs the grace space to implement their plans rather than constant interference with every emerging ideological issue or self-righteous whim.

Because civilised society has always been decided by robust face-to-face relationships enabling local face-to-face collaboration, global face-to-face collaboration, and faith, the centralised DeFi/ web3/FIAT crypto strategies would destroy this fabric of society. So, rather than being decentralised, they do, in fact, trap and atomise people in their individual bubbles without any central power to support society, values etc, let alone laws, leaving the marginalised and disadvantaged a growing group vs the elitists. We can see this polarisation and nativity now. There is no evidence to suggest unless people get completely addicted and nullified by tech, that this will end well.

These self-proclaimed “decentralised systems” are exactly the opposite. They open nations up to total central control and require governance, ethics etc. With less face-to-face communication and atomisation, we will destroy relationships and not value others. Why should we value others in this impoverished tech, intermediator model? We are seeing a truly terrifying outcome now.

Tech organisations such as Amazon vs the traditional alternatives such as GM employ about 10% of the employees for the same revenue. Most of the tech roles are on/near the minimum wage.

In the early stages of the tech business development, their valuations get hyped because of the potential, when in fact, these businesses because they have little differentiation, not much that is difficult to replicate, so may dominate for a while, but don't engender profits. So low/transactional profits low and negligible loyalty are a land grab/race to the bottom.

Emerging tech industries are not compelling or innovative. Conversely, traditional industries historically innovated to create new high-value industries that in turn employed highly skilled as well as a range of support roles at all levels in manufacturing etc. So, these tech businesses largely trash their own market and parasitically rely on and sap margins from their host: productive industries, only to be replaced by another for diminishing returns. This pattern is destroying value, trashing the fabric of society, reducing opportunities, so these tech industries don't suit people or society and will make us poorer!

The economic rental they impose on productive industries through imposing large margins as intermediators of the productive industry online sale controls the pricing for the productive parts of the economy, driving out competition. Addiction to buying online fashion creates cheap clothes, of which 60% go straight into landfill. Landfill because the buzz dopamine addiction is in the anticipation of the purchase, NOT the owning or wearing of the item!

So, ideas that don't fit with the ideological ideals of these powerful tech business models get silenced, and when attempts to improve, say by addicting customers less, commercially, their tech sales/control of attention fall off a cliff.

Controlling attention is all about keeping addiction. In so doing, social media controls free speech that may relate to these tech downsides. So, people who disagree with the associated echo chamber get taken off listings, thereby conveniently reducing their influence on social platforms.

Processes should not be designed for process exceptions, yet when it comes to education in schools, gender dysphoria, trans, or such activism etc, tars all the same pupils with the same brush to normalise abnormal behaviour.

The norm and crux of society are heterosexual relationships. So, heterosexuality should be the crux, alongside an acceptance of and love for others, such as LGBT etc. Teaching the normality of the LGBT+ exceptions create a contagion of gender dysphoria in girls. This confusion destroys the basis of the male and female role and role models in society. The nuclear family is the basis of our civilised society, the development of our kids and our identity.

So, gender dysphoria should be treated by exception rather than lauding self-identity and taking extreme offence at disagreement. Self-defined identity is the infantile behaviour and stage a two-year-old goes through and is narcissistic. One's sanity is defined by external relationships with others, so everything cannot be self-defined and become a fact, as we wish at whim and impose on others. Most of our ideas are wrong, and if they don't survive contact with reality, they should be discussed and or reconsidered.

So why the attack on identity? Woke arose in the 1950s, and like BLM and other neo-Marxist woke organisations, in part, neo-Marxism is focused on destroying the nuclear family, as it's the crux of society. In the same way, the neo-Marxism *modus operandi* is to infiltrate organisations top down and then create conflict in the ranks by focusing on victimhood and corruption by evoking emotive falsehoods to gain control by infiltrating, controlling and constantly redefining the narrative. It's a power grab for those who feel entitled, which drives fear and silence.

Silence, for example. The following week after George Floyd was killed by police, a white man was killed using the same police restraint technique, yet his death did not make the news.

Corrupting identity in this way successfully corrupts the morality and backbone of society, reducing the will of people to fight and speak up for fear of being terrorised and cancelled as antitrans, anti-racist etc, labels by exception don't follow. So whilst very nasty, they are just nonsense.

Furthermore, the false accusations of racist anti-trans get so overused they lead to cry wolf and real issues get covered up.

The pandemic and shutdowns re-established the importance of relationships with our neighbours, reconnecting people, using less, getting fit, and getting back to basics.

The Germans were warned by Trump that sourcing gas from Russia would be bad news, yet China is in the blind spot and remains a far bigger threat.

The WEF view is to professionalise this anti-democratic trap by reducing people's influence by: people owning nothing, EVs replacing industries, reducing population, people being able to afford EVs, less income less industry, ESG/nitrogen targets making Dutch farming unviable, centrally owned assets and controlled in centralised intermediator models, without national government, cultural tyranny of ESG targets imposed by big investors, crypto and blockchain voting systems redefining countries regardless of boundaries and undermine governments.

"Productive" business confidence is low.

Ignoring the cause and the "TOP 7"

The pandemic gave us a glimpse of what good looked like: we reconnected with neighbours and found peace. During the pandemic, we conducted extensive research with CEOs and thought leaders to create a summary of the main learnings to shift away from growth and domination to these top 7 subsets or factors that constitute profit and relationships:

1. create work for all

2. increase prices and profit to reduce consumption

3. protect nature and re-use to save the planet

4. people relationships – reduce suicide

5. reduce online technology – regain self, less addiction

6. support local and community

7. intimacy: acknowledge and value everyone's view

And 8. Extract from misaligned regimes such as China and Russia, or those that you can't say no to because, by default, they will take over.

The common denominator for success, the tougher things get, is to put others first. This is also the first commandment – that is the crux of love. Love is also the only indestructible emotion.

The key is to support relationships with Putin, or do you think you know better? The cause of Putin going off track and disconnecting from what is considered acceptable was his isolation during the shutdown. Yet whilst it's convenient to label Putin as a tyrant, the Western invasion of Iraq was equally if not more flawed than his justification for invading Ukraine: the encroachment of NATO on his borders, the same cause of the Cuban missile crisis, and the West's focus on neo-Marxism.

During this troubled time, China has sat on the side-lines with their propaganda, sharing the choice Order vs Chaos, which to the weak-minded (in the context of the West falling to pieces) looks increasingly attractive. As this is appealing, it's also propaganda, so flawed: the Chinese system is 100% based upon ideology and self-preservation of an individual Xi. An individual who promotes themselves to such a God-like position is hardly a proven workable model, so needs, like any tyranny, to be viciously controlled using technology to keep the fear, social scoring and regime in China.

Social scoring seems to be creeping into Europe, starting In Italy via an app in 2022. Social scoring is the beginning of the end of privacy and the beginning of hard communism, but obvious technological means of gaining control of the population as per China under the banner of "Encouraging good behaviour!".

The solution to realign is a spiritual one.

The solution to realign is an inner game, one of grace and love. The alternative to grace and love is anger, a lack of forgiveness and self-righteousness. We are seeing this anger and escalation now, so the tech to slow it down using social scoring and to silence rather than fix the root cause, which is spiritual, is tempting but will screw up the best high-integrity people.

Culturally if you can view business as a fight, artificial intelligence ultimately is the ultimate fighting tool to fight it out between companies and an excellent strategy to win, but at what cost?

China CCP views business as an attack tool and has no interest in collaboration or supporting local and national interests, so artificial intelligence is a good fit.

Artificial intelligence creates a lot of reworks, and we can't keep up either, so it destroys people's bandwidth or patience to engage, a perfect silencing tool.

Those who care sufficiently to speak out are being de-selected by artificial intelligence. They are not being listened to. So why speak up at all?

Imran Kahn suggests that there is a materialistic route or spiritual route. The spiritual route is the tougher one but driven by a far deeper sense of purpose beyond us. There are a lot of commonalities between Islam and Christianity as well as wide divisions. The common link is the Bible. Islam was mentioned in the Bible and predicted to be problematic.

From 200 company turnarounds, the common denominator to putting companies back together is grace and love, or else you can't move forward – because the alternative is that executives are stuck in their pride.

So, there is a basic choice to find the courage to take a spiritual-led solution to get back in control and reconnect with our purpose. We need to either learn why we are on earth or carry on proudly trashing the earth, creating weird ideologies, using technology to control, and thinking we know better to fuel the tyranny.

The choice is one of madness – carrying on doing the same things expecting a different result or switching?

Technology, therefore, largely doesn't fit the spiritual model because you can't rewire it with God-like characteristics without implanting: ideology; silencing; weird, self-defined, self-serving ethics, and the endless amorality based upon redefining identity.

People are the connection and manifestation of Jesus on earth. It is through Christians that Jesus is a living God. Jesus is love, He lives in Christians and even enables us to love our enemies.

Social media technology claims to be social when the reality is the opposite. It atomises people and undermines their uniqueness and quality of communication. It is ideal fuel for neo-Marxist victim-centric rules and attacking dialogue. These attacks hold people in fear, reducing our ability and interest to communicate because the ideology becomes predominant, preventing getting to the interesting part, getting to know the person underneath.

Our well-being is decided by being acknowledged and valued for who we are, as we are, warts and all.

So, attempting to create a utopian man means destroying traditions and institutions, replacing hatred and emotion, or most recently blindly following science. This is odd when love is the only source of well-being and indestructible emotion. So, the neo-Marxist/WEF/ESG/EDI utopian person is just flawed ideology that leads to tyranny and parasitic trashing of those aspects that create wealth; institutions; value in people; relationships; profit; grace and love etc.

So, it's important we discover our humanity and get back to creating productive industries and productive work for all.

The antithesis is what we face now: industries that are parasitic, parasitically destroying productive industries. The Big 4 follow these gravy trains yet create a worthless future themselves.

These industries can ONLY survive when based upon hysteria keeping alive their ideology.

Fintech, crypto, web3 etc all follow this pattern. Parasitic because they do nothing new, are not compelling and don't enhance the overall benefit to society, relationships, or profit.

It's the hysteria that is maintained by the social media echo chambers, which maintains their flawed self-perpetuating destructive ideologies. It's only a good crisis that roots this out because there is nowhere to hide in a good crisis, so it forces everything into the light.

1. Destroy the traditions and institutions (to rebuild with a disregard for the past).

2. Notion of utopian man, socialisation.

3. Attack nuclear family: (the construct of society).

4. Denial for human reciprocity (things need argument).

5. Fear – based upon ideology - Straight good and evil, (replaces grace and love now creating

polarisation and no time for others).

6. Individual and international socialism – centralised usurping governments WTO, big tech,

Big 4 etc.

7. Science vs ought, science says you ought (blindly abdicate responsibility by following science).

8. Hate landowning and the rich, (yet the post-modernists don't care about the poor either).

9. Too many people create a human policy (dismantle and weaken those who refuse to accept

what they say).

10. ESG-type ideology outweighs commercial sensibility, (Dutch farmers' nitrogen targets imposed

by the WEF!).

11. The WEF ideology is flawed and not working (so speed it up).

The impact of this is not to love your neighbour as yourself. So, there is no reciprocity, which is the basis of society.

The disintegration of people who care by being marginalised, to the point of not caring, so not listened to, so no point in speaking up, no effective dialogue undermining relationships and integrity.

It is the people who care that steer the ship in a good and sustainable direction.

The WEF/ tech-woke type mess creates the perfect set of circumstances to manipulate the masses.

• Escalating rework – excessive errors and mental illness.

• Impatience, selfishness, too busy, no dialogue, no time for each other.

• Speaking to ideology – not person – not interesting – lost source of value in communication.

• Hijacked reality with lies, “governments fuel inflation”, “crypto solves inflation!”.

• Insular perspectives without good information – leaving the common man with no idea

what is happening, or worse, caught in the neo-Marxist stupor or immobilised and addicted to tech.

• Young people addicted to tech and trivia, behaviour control – addicted and taken out (UK education now leaves out Churchill!).

• Manipulation, isolation, social media because key media is owned by big tech.

• Shortcuts and corrupt strategies

◦ Crypto, social scoring

◦ WEF and big tech-owning land, including China

◦ China propaganda Chaos vs Order

◦ Simpler big tech intermediator aligned products – EVs – less work

◦ Simpler products – lease model not owned too expensive to own

◦ Digital diversion of liquidity away from FIAT currency

◦ Interest rates trashing asset prices – can't own property

◦ Parasitic tech trashing margins of productive industries, booking.com types, Google EVs etc

◦ German business has gone too far with China dig roots in deeper

• Communication skills are reducing; less empathy is causing people to fall out with each other.

• Addicted to tech – can't think properly, short attention span, no counsel – immobilise our young, our young should be getting angry and revolt, but instead they become mentally ill?

• An escalating challenge with no learning – really stuck on your own hamster wheel.

The above creates stress because it dislocates people's actions and beliefs, creating a loss of self and identity and purpose. This is exacerbated by real financial and societal difficulties.

Communication skills are the basis of relationships.

Being stuck on technology, children don't know what is happening around them. Post-modern idealism has stuck them in fear of speaking out, undermining their sense of identity, let alone developing their communication skills.

Therefore, timeless Biblical principles are now so important. Jesus died on the cross to have a direct relationship with us – the crux of profit is keeping relationships:

• love God, love other people – foundations of ethics/strategy

• do you know better than God? (vs. Pride and Hubris)

• we are uniquely made in God's image (value the individual as they are)

• if you are without sin, cast the first stone (recognise you are no better)

• look after the poor, the meek shall inherit the earth (material aspirations are shot)

• gift from God to appreciate what you have – (antitheses is more, more is a curse)

• giving is more blessed than receiving, (material aspirations are shot)

• bring things into the light – (or else silence fosters evil)

• ethics flow from the freedom found in Christ – (freedom, relationships, and accountability)

Yet for the proud, the growth and domination mantras continue, leaving people to look after themselves, cut off dialogue, and lose sight of what they realised was beautiful, good, and true from the pandemic.

Silencing

Misaligned actions are subtle – such as social media. It is a trap. Social scoring has arrived informally in terms of self-defined post-modern ethics. Technology is and will make this increasingly formal, rewarding those for good behaviour and cancelling those who fall out of line, as well as being cancelled for challenging the government/national vaccine policy online, regardless of its merit.

You won't learn or fix the issue until people admit they are willing to explore whether they have done something wrong, so free speech and bringing underlying thoughts into the open is vital.

Social media is dividing people into echo chambers to avoid conflict and healthy debate around the big issues, reinforcing their self-made hysteria language that keeps people stuck in their echo chambers, believing the untruths.

Tough times are the acid test because this nonsense comes out into the light and doesn't pass its first test of sanity on open contact. Yet it has been promulgated in the echo chamber for years, so people will naturally and automatically reject a challenge without thought.

The proud can't learn unless they are willing to explore and adjust or admit that they might be wrong. Increasingly people become less interested in what others think, so they resort to trashing society to manipulate their own agenda without valuing the source or recognising that we all see the world differently.

An extended period of falling in line with ideology and hysteria is hugely costly.

Productive communication and meeting with a diverse, courageous, open-minded group prevent that.

A totalitarian state only works upon lies because it's too complicated and inappropriate to define rigid rules which are any good. The audience is not homogeneous either. Freedom with proper boundaries is far more right alongside humility. From these perspectives, you realise your reliance on others. This increases self-awareness and self-fixing and is a huge burden lifted.

It is hell and difficult both ways, either:

• being honest to face the facts,

or

• distorting or removing the truth

Tough decisions. You can't win against those like Putin or Xi, who you can't say no to, and/or if you are an employee who disagrees with post-modernism corrupting Virgin Airways with their identity-free dress code!

Big 3 investors, including Blackrock and Vanguard, dictate their ESG ideological policies and buy up property en masse to control key assets. Media is owned and funded by the US Government, Bill Gates, and big tech.

The intention of these large organisations is to nullify power corporately using their ideology at board level and undermine boards and owners: such as reducing the viability of farming in Holland due to the WEF influencing the Dutch Government to introduce nitrogen targets to reduce the influence of landowners and Dutch farmers. Then Bill Gates and retail banks that pertain to the WEF ideology can invest in and buy this land on the cheap.

Rather than this WEF being seen as fascist, this takeover and emerging tyrannical control is all masked in a caring/liberal type of way.

No work, no money alongside social scoring – no control, it's clear the underlying assumptions are flawed too: no one in this utopia will pay or be accountable for the universal minimum wage.

So, we are stuck and can't stop. Tech is duping and addicting, successfully trashing society relationships and profit to gain a parasitic type of control of productive industries without the rubber hitting the road. Consumers can't stop because they are addicted to the tech routes to information and market.

China has an intentional strategy to get control of other countries through loan – own model.

Whilst suffering serious losses, the belt and road initiative is gaining traction in Germany and Italy, leaving these nations open to situations that they can't reverse.

Resolving the heart of the matter is a systemic spiritual issue. So, these post-modernists systemically trash and mock faith too, as being non-exclusive.

Blame is the sign of victimhood, scapegoat, and terror, the crux of post-modern ideology, even blaming the Government despite being incredibly generous during the pandemic.

Left-wing ideology that doesn't value the individual squanders innovation and leaves people poor.

Rise of pointless businesses fuelled by hysteria and whims, crypto boasting that gilts are toast, governments intentionally driving inflation and reducing the value of your money - as is crypto – is a solution, or justifying a blockchain solution based upon a flawed premise without addressing the root cause nor proposing a better solution – that defines reckless: Cutting through the lies is key to getting the framing right – crypto is gambling and addictive.

Identifying patterns, such as a Ponzi scheme, means that those who started buying at the beginning have been able to acquire large amounts of bitcoin and sell them for a profit, but those joining now are out of luck, and if the last person doesn't buy for more the value reverts to zero. This hamster wheel is fuelled by the hysteria of the echo chamber.

Crypto is almost all speculation, no cash flow or utility, just an accounting entry for a non-existent asset, so it's unstable, effectively, at best, unlicensed gambling.

Marginalised people will resort to violence if they are unable to speak. Although if they do, their dialogue is wasted online because social media algorithms make sure no one sees their communication, so shutting down their “subversive views” online.

Social media, as an act of their own self-preservation, will limit views by removing them from other users' feeds for fear of an ever-escalating discussion that might get out of control.

Lies and propaganda of LGBT, the menticide of controlling language, redefining language, alongside no truths, and not valuing truth makes good people angry and dupes those who don’t care.

Issues to resolve.

• Find a way to measure strategy vs profit and relationships.

• Without spirituality, a higher purpose – there is NO robust purpose or meaning, you can't define your purpose, you are not qualified. 😊

• No society has ever survived without faith, it's the common denominator that creates the structure, both seen and unseen.

• No one is talking about faith, so the structure post-modernists are trashing stays unseen.

• Prevent people from controlling dialogue or redefining dialogue – dialogue controls us.

• Stop doing stupid things such as the metaverse that is an attempt to create the ultimate dystopia.

◦ Understand the underlying thinking, maybe: when communication breaks down completely, and we resort to our avatar communicating on our behalf.

• Accept that you can't define your own ethics without a higher meaning and purpose; how can you decide ethically what is right or wrong?

• Slow down and think – or else we are captured on an accelerating wave of fear, fuelled by flawed ideology.

• Scrap echo chambers, listen and speak to people you don't want to or who have different views.

• Get off social media.

• Invest in professional development and overcome your pride and denial, or else the epicentre is blown around based on mass ideology/whims.

• If something doesn't sound right, speak up until you get a conclusion, that is the trigger for critical thinking.

• Take tough decisions where you are trapped by a maligned force, China or Russia or social media etc.

• Technology is based upon ideology, so unlikely to fix the big issues, protect the natural environment, well-being, and nature first and reduce focus on more – drives the curse of consumption.

• Focus on profit and relationships, not growth and domination.

• Stop stupid ideology woke, post-modernism – that's professionalising mental illness.

• Face the facts, strive to reach the truth, see, and upskill to become aware of distorted and cutting factors that corrupt the truth.

• Relentlessly searching out the truth makes strategy simple.

• Watch out for the big agendas, study the dialogue and drivers of big tech/WEF, ESG EDI taking control of land.

• Get to know the person, cut through the ideology – value what they say and realise the beauty of people.

• Social media – manipulates and feeds your echo chamber – have arguments.

• Avoid and become aware of extremism – no counsel around you, meet with peers.

• Groups fracturing NATO, Turkey G7 and Russia love thy enemies, support dialogue, value individual views, get views out in the open.

• Women's roles are different, far better at managing risk as part of a team.

• Recognise lies; there is no diversity in the colour of our skins or sexuality, it's a falsehood.

• Life is unfair: the main unfairness and driver of success is intelligence; the successful have a responsibility to look after the poor.

• Break the cycle get off tech and social media, get back in control of your thinking and yourself.

• Spiritual connection – get ethics right.

• Re-establish relationships, care, get to know, set up logos as the seedbed for jazz band, means of steering your ship.

• Value thinking and critical thinking, trigger that doesn't make sense.

• Resolve flawed dialogue, root out ideology, become aware of the impact and blind spots, avoid value statements – creates ideology.

Reset the structure and pillars:

• Spiritual moral compass and logos, meetings, connection, and intimate relationships.

• Dialogue grace and love thy enemies, love fixes relationships – put others first.

• Become aware of flawed dialogue and where your dialogue is silencing or creating blind spots.

Enable the right people to thrive

• Avoid the Boris Johnson effect. Critics become overrun by self-righteous anger based upon generalisations party gate, liar etc, practice forgiveness to prevent anger and support grace.

• Understand we are all flawed.

• Notice that any action is always flawed – the alternative is to shut down new action!

• Address the big issues, Brexit, good team, fiscal control etc.

• Avoid getting drawn into generalisations, get specific. Negative generalisations are always false.

• Make hard calls and stick to them. Don't get usurped by cowards. Stick to the plan and value rigour.

• Avoid following populist agendas that ignore sensible financial or ethical rigour, e.g., Liz Truss

plans to spend her way out of debt and cancel national insurance rise to fund social care.

• Become aware of flawed thinking – get control of your dialogue and inward challenges – specifics no generals.

• Focus on the top 7.

• Find where risk is uncontrolled, China and cyber, and crypto impact/diversion of funds on FIAT liquidity.

• Practice grace and love, forgiveness – or else you are faced with anger, cancelling and hatred; that's hell for the canceller.

• Get back to basics, nature, appreciate the beauty around us and the beauty in your peers. Understand and protect nature – nature is the majesty of Jesus.

Framing and diatribe create blind spots and justify the status quo.

Take the fall of FTX, (quoted from Kelvin Lowes work) quote starts:

The *New York* magazine reports:

"One of the crucial prongs of Sam Bankman-Fried's big con was his co-opting of the press, which turned out to be very easy for him. Manipulating journalists was a cinch—so few of us know how to write about anything having to do with crypto—and he threw money all over an industry that's notoriously cash-starved... SBF knew how to play reporters. He gave a good quote, and he didn't sell too hard, acting as though he too were a bit sceptical of the whole crypto thing. That posture stands in contrast to how most evangelizing crypto cranks speak to reporters. As Bloomberg's Matt Levine (one of SBF's favourites) wrote, "A wild-eyed crypto true believer is not the person to operate an exchange. "Reporters covering crypto were desperate for someone in that world who was not a craven scammer.

"People in the media wanted this to be the guy in crypto who is actually doing the right thing, and he was supposedly allied with regulators, supposedly on the right side of this," says Davis Richardson, a Web3 media consultant, "but he's actually doing the same shit that everybody else is doing in the industry."

None of this was true, of course.

And he invested in Semafor, which reported that he had tried to buy his favourite bloggers, Matthew Yglesias and Nate Silver, for a Substack-like publication he wanted to start.

"If he had any relationship with me, it was purely parasocial," says Yglesias..."

But SBF does have a particular fondness for Yglesias types, those Twitterholic, progressive-lite pundits who looked fondly on (and could translate to readers) the high-minded philosophy—effective altruism — that supposedly was the impetus for his billion-dollar fortune. As Yglesias wrote in May, SBF's "cryptocurrency businesses just exist to finance effective altruism." While Bloomberg and Fortune provided him with a reputation for being a quirky League of Legends–playing, dishevelled kind of Buffett, it was the wonkier voices in progressive media who fuelled the narrative that he was also an exceptionally decent billionaire. The Nation's Jeet Heer in mid-November slammed Yglesias, accusing him of "whitewashing a transparent conman." Zack Seward, a contributing editor at *CoinDesk*—the industry publication that first brought the house down on SBF with Ian Allison's report exposing the lunacy of SBF's balance sheets—said that, above all, SBF simply "craved mainstream legitimacy through whatever means necessary."

And so, this sociopath basically manipulated credulous journalists who built him up to be some sort of crypto messiah.

Although "that which the press builds up, the press tears down", how does this comfort investors who have lost everything?

Decision-making and strategy depend upon effective communication. Effective communication is the seedbed of relationships, and relationships are the seedbed of enduring profits. Measure actions vs profit and relationships, not growth and domination.

Watch echo chambers and glasshouses such as crypto web3 that don't hold water, are pointless, parasitic, and collapse. The underlying hysteria undermines the integrity of your best people and your strategy.

People who complain are NOT toxic; they care. They are what diversity looks like; understand and value them.

Leverage women as our beneficial adversaries. Test ideas vs your broader network.

Let women be women, manage risk and speak up. They care about people more too.

Reset relationships, set up logos, intimate connections, get to know each other, and show empathy, humour, and banter.

Kick out ideology that leads to tyranny.

If something doesn't sound right, challenge it, and don't back down until you get an answer that makes sense.

Listen and don't interrupt others when they are speaking.

Learn to marvel and appreciate what we have – this is a spiritual issue.

Establish robust and intimate relationships, develop logos, give people freedom, check quality.

of ideas and dependencies in reviews, use meetings to check the sanity of ideas and alignment, leave accountability with the individual, leaders serve, directors direct, set boundaries or "the what", listen to and acknowledge people for who they are, take an interest in understanding and valuing different views. Voila. Laugh and have fun.

Watch out for silence, what is NOT being said.

Take EDI /ESG – really? What's the impact?

If you have rolled out an EDI strategy, what should your staff say to you, maybe:

• This ideology is creating tyranny.

• You don't value what I think.

• Those values are simplistic and silence people.

• The anti-bias training is like the CCP.

• The best people are silenced.

• The wrong people are thriving.

• People are committing suicide.

Yet what are they saying – NOTHING?

The critical capability in your company turnaround is to define your core business and cut it back to a defendable core.

Then from that position of clarity and confidence, realign your commercials and team to maintain a step change in relationships and profit.

I share these underlying learnings from our work in progress and the last 20 years leading in company turnarounds to steer you towards keeping you and your business protected, strong and healthy.

I also share this to check my own thinking and compare notes with my trusted network.

So, what does good look like NOW? It looks like a Jazz Band.

BUT.

Their plan is working perfectly.

The problem is that it's not your plan, nor is it in your best interests.

And worse, you have no clue, so you have Bob'all chance of knowing how to get a grip.

As a production superintendent at 20, I remember looking at what was on the month's production schedule vs what my team was working on... about 70% of their activity wasn't on that month's revenue target!

That's costly. It always resulted in excessive overtime at month end to achieve "the required or invoice-able" sales.

Then it was relatively easy to fix issues: there was far more clarity, and dialogue was much more productive.

Yet, now have you noticed dialogue is getting increasingly weird and difficult.

For example, “Woke is a good thing.” 60% of women agree with this, so do 70% of millennials.

But what if the underlying intention of the “Woke is a good thing” dialogue was to recklessly pervert your own interests?

“Woke is a good thing” illustrates a serious problem and one that is very hard to fix.

It's not about being right or wrong either. It's about us being able to maintain productive dialogue.

In particular, the ability to maintain productive dialogue when things get tough.

Take Silicon Valley, Xi in China, and Putin. Their primary interest is to protect their power and wealth.

Yet you would do the same if you were Putin?

You probably think an association with Putin is not relevant here. Putin is an evil tyrant. I am not an evil tyrant.

So how might you go about protecting your power and wealth?

And how well are these powerful interests aligned with your own strategies, or you getting duped and dragged along?

This is how it happens.

Take the World Economic Forum; WEF: collective organisations such as WEF, create mantras and ideology.

The WEF's current killer mantra is “climate emergency”.

It doesn't matter what the ideology is; just like woke ideology, it needs to be emotive.

Then.

Do you recall this started in the 1970s, at which point, this mantra was "climate change" which then shifted to "global warming"?

But "global warming", unlike Putin, sounds quite nice?

So, to ratchet up the ante, the WEF has now given us the mantra of “climate emergency”.

The WEF quorum consists of the Big 4, Big Tech, and most global leaders.

So, what is the cost of the climate emergency mantra or those ideologies/mantras such as EDI etc?

These ideologies work together in their joint missions to force an agenda to get control and, in parallel, silence healthy dialogue.

Take Judith Curry.

She is a formidable climate scientist and former academic.

She correctly challenged the "climate emergency" mantra so consistent with the underlying intent of EDI, which is rife in academia – and was forced out of academia.

So, what are the emotive elements of your organisational dialogue?

Trust me on this.

You have no clue.

We have done work training and engaging with the 3 of the Big 4 and many corporates – the answer is they are totally unaware of their internal dialogue.

They have no idea what to look for, their dialogue is so familiar to them.

So off the radar and unchallenged.

Why don’t they have any idea? Because this dialogue is in their blind spot.

Blind spots cause accidents and surprises.

So, you have no clue.

So, if you can’t see them, let alone speak up against these powerful internal mantras, is it any wonder that the Big 4 auditing outcome gets corrupted?

It’s the bright young graduates who do that work. They are completely unable to raise difficult issues that come out of audits.

How does this woke ideology do that?

Wokies, under their evolving, emotive banner - EDI, LGBT+, CRT, intersectionality, queer theory etc - attack by taking emotive ideology around emotive subjects - race, sexuality, and identity – to associate unrelated things.

Take football in Qatar. You are forced to “take both football and LGBT+ on a take-both-or- none basis” such as football and LGBTI One Love ideology.

Yet there is much to be learned from the Qataris. What do you know that is great about the Qataris?

Yet we disguise such ideology under EDI - notice the outcome is exactly the opposite.

It’s far more damaging because if you screw up someone with lies on one axis - you screw them up and corrupt them on all of their axes!

That's why letting adverse audit results go is so damaging: what happens at the next audit?

So like any subversive lies-based system, it has to evolve.

Evolving unchallenged lies and propaganda creates menticide which is why football fans, formerly renowned for violence, are silent.

Lies have to evolve because as the collapse of Crypto Asset lies show, lies don't stand scrutiny.

They create a mind virus.

Yet crypto enthusiasts see the crypto collapse as a reason to accelerate decentralised finance.

Echo chambers fuel and accelerate and entrench such madness.

So silencing a group such as football fans known for violence is quite an achievement!?

One of the underlying attack tools is intersectionality. This is where every point of contact is weaponised by nominating a victim and oppressor, the tool of neo-Marxism.

Survivors will avoid manipulation.

But what are the skills to avoid getting duped?

The mind virus psychology legitimises lies or encourages generalisations from isolated scenarios.

This beautifully inflames marginal issues and creates real fear.

Generalisations create conflict and silence.

Take the generals of "Brexit" and "Remain". Is that dialogue concluded?

Yet if you know how to get past the generals of "Brexit" and "Remain", the specific differences are small, and if you know how, a proper conversation can take place.

Yet the polarisation around "Brexit" and "Remain" – has just become entrenched since 2016, the decision being made?

So, can you have a "Brexit" and "Remain" discussion with your peers?

Or have you given up?

What else have you given up discussing, and/or who else have you fallen out with?

The longer this goes on, hard habits form. After 66 days of doing the same thing, habits become permanent.

So, with less dialogue during the shutdown, alongside the fear and practising silence, you will start to believe your own thoughts unchecked!

Generals are more tempting the tougher things get. It's the way the self-reliant cope when faced with impossible situations because the issue is down to YOU to resolve?

So, you have to normalise your responsibility away, e.g. "Putin is an evil tyrant".

The mantra of "climate emergency" has been driven by an underlying political agenda or WEF to protect their power.

The WEF interests are to create a world government and undermine capitalism or local wealth creation. Local wealth creation is a threat to WEF's power.

They maintain power by choosing global warming as their emotive mantra/issue to get around and create, manipulate and fire up their activists away from their own interests.

Yet, as Judith Curry notes, many natural disasters would have tragically happened anyway.

Yet notice how evidence for this mantra is selectively aligned with climate disaster?

Leaving the WEF with no interest in considering the option of these events being part of natural climate variability.

So, what's next? Greta created a cult, a psychological problem of hopelessness, leaving some of our kids suicidal and fearful of having their own children.

It's tragic because our kids can't filter this. They are overwhelmed without thinking.

Depressed kids are a big problem: what will this mean for the next generation?

All for the "climate emergency" political agenda to blindly get rid of fossil fuels?

So, what is a more sensible hypothesis and psychology?

What is the biggest risk?

Let's go back to the silenced Judith Curry. This is her robust observation and hypotheses.

Moving too fast from net zero using wind and solar is the biggest risk to the environment right now.

It makes complete sense, because that solar and wind money should be invested in preparing economies for local natural disasters, NOT an immediate switch to wind and solar.

In this way, mitigating damage to our economies to pay for protection of the environment.

So, moving immediately to wind and solar will cripple our own and developing economies and completely cripple any move to NetZero!

In the meantime, fossil fuels are ALSO needed in a developing world to enable this transition and develop and implement the solutions over the next 30 years.

So, the psychology of the "emergency" mantras forces the focus on making the link with our activity and extreme weather events.

Take ice caps melting, she observes are probably as much due to subsea volcanic activity than C02 and that needs to be planned for and impact mitigated too.

So, EDI etc, Judith has been cancelled by the academic world.

So, who have you cancelled?

Or in fact, who feels liberated in your company?

Or how much courage does it take to speak up against the same type of falsehoods in your organisation?

Or don't you see or recognise your falsehoods?

The weather was worse and more disastrous in the 1930s than now, and yes, the last 5-year cycle has been bad too.

The earth is NOT fragile either, it's a perfectly created impossibility, that could only have been created by a theological force.

Let’s take a broader check:

From a Biblical perspective, the environment is looked after by God NOT us, so, thinking we know how to fix the infinite complexities of nature is absurd.

The more scientists investigate the universe they realise the LESS they know, not the other way around. Creation is perfect down to atomic level, driven by beautifully perfect interrelationships.

So, our responsibility is to look after the environment, not fix it, we simply can't.

So building our economies in order to be able to invest to protect our natural systems and defences is sensible.

Yes, sea level rises are a problem, but here is a blind spot, we also need more water than global warming creates.

Climate events we witness now are mainly inevitable, and the inertia of our previous actions are largely impossible to stop.

So, global warming is not necessarily bad.

We also will need more energy in the future.

So, the focus should be to protect emerging economies from climate events to grow their economies in order to be able to afford the mitigation actions to protect the planet over the next 20-30 years, not fix it!

The C02 removal tech to remove C02 from the environment are truly ridiculous.

In the meantime, defining burning wood and dung as being sustainable is absurd, because the pollution is very bad for the environment.

Do you see the flawed parallel of EDI/LGBT+ ideology being good for our well-being?

These ideologies are tools of control.

The Jazz band Metaphor:

What does good look like? It looks like a Jazz Band, a group of musicians with an intimate relationship that allows them to create logos – a strong sense of what is right and wrong, read the audience and immediately take new action. By taking new action, they will learn, transform their own capability, endure, and remain on track.

People must re-experience what good looks like, and playing in jazz band for themselves, people will learn nothing if told so cannot be told.

My Love to you all.

To appreciate the solution, you must experience or re discover what good or experience winningthinking® to experience that re-connection with what is true, good, and beautiful.

Cheers Tom

THE SOLUTION

From our research these are the top 7 shifts including avoiding nefarious powers in your supply chains such as China.

To take the new actions to see and read the strategy, we have found you must take new perspectives to inform and take your own new self-identified actions, as well as upskill to avoid falling into the blind spots and ideological traps that corrupt people, critical thinking, and strategy. A Winningthinking® masterclass enables that.

Our future depends upon building a better world:

The 7 permanent shifts for CEOs to action to avoid the race - bottom:

1. Create Work **for all levels: protect Employment**
2. Increase Prices **and Profit: reduce consumption**
3. Re-Use: **reduce consumption, protect the planet**
4. People Relationships: **reduce suicide *1 etc.**
5. Reduce Online / Tech; **less addiction regain self *1**
6. Support Local and Community: **enable mutual success**
7. Intimacy: **Acknowledge and Value Everyone's View**

& 8.. exit China etc.

Underlying focus for success - "Put Others First"

When the process of taking new actions and becoming aware of the traps and teams restart thinking you are back on track to realise logos (the source of freedom and ethics), individual accountability and intimacy with your peers, that enables you to realise beauty, truth and what is good.

Strategy is very easy from that point therein and steers itself with minimal external intervention.

References

Johnson, S., Acemoglu, D. (2023), *Power and Progress: Our Thousand-Year Struggle Over Technology and Prosperity.* Basic Books. A great book, and a primary source of research for the feudal aspects.

Michael Nehls, MD, PhD (2023), The indoctrinated Brain: How to successfully Fend off the Global Attack on Your Mental Freedom. Skyhorse Publishing. A great book, and a primary source of research for the feudal aspects.

www.ingramcontent.com/pod-product-compliance
Lightning Source LLC
LaVergne TN
LVHW052111210926
848482LV00025B/541
9781739088057